uch like pools in tropical forests "like the blood of the trees," and probably death to any white man, who, by the pangs of thirst, would drink thereof. How I was longing now to get back into more cultiva s, for where this man was taking me I knew not, and I was completely in his power. How ter long wandering, and many fearsome doubts, with many encouraging instructions to n," we actually did emerge at last to where Cassava was growing, though still I saw no rk that I could recognize, and still I followed on in the wake of that tall, black Mohammedan sun which had been behind clouds most of the time, when we were in the dark shades of the now came out in relentless, pitiless fierceness, which tended to increase my raging and make the constant attacks of heavy sweating to which I am still subject, almost mo could bear. But we were getting back to civilization, this man, in whom I had had lit no confidence was taking me to the Lagos road right enough; only what a distance it emed, and I was now almost mad with thirst. At last he paused, and I did see not far away ooked like the back of the Leper Laboratories, and also a fairly large house, standing in its ounds, which indicated on a board, belonged to the Salvation Army; and having got a youth who spoke English, I was able to express my very sincere thanks to my most -worthy and competent guide, not without some pangs of conscience at the though much I had mistrusted him; but I believe he had been coming that way anyhow, an st only allowed me to patter along behind him, which was as well, for in any cas o coin to offer him, and I doubt if he would have accepted it if I had. (The man in Alexa id not.) It was now more than 5 hours since I had left Miss Tracey and I had my feet all the time, so that I was now feeling about done for, and still that ragin So seeing a kind of rough seat, near where a woman was picking over cashew nut crawled slowly towards it, and sat down, but I could get no information whatever out of oman, and now, added to my other miseries, I began to have internal pains, whic nded a speedy retirement into the bush. Then it occurred to my obtuse brain tha e was a place on God's Earth, where "the cup of cold water" could not be refused, that

BUTTERFLIES
AND LATE LOVES

the sequel to *Love Among the Butterflies*

BUTTERFLIES AND LATE LOVES

The further Travels and Adventures of a Victorian Lady

MARGARET FOUNTAINE

Edited by W. F. Cater

SALEM HOUSE PUBLISHERS
Topsfield, Massachusetts

First published in the United States by Salem House Publishers, 1987.
462 Boston Street, Topsfield, MA 01983.

LIBRARY OF CONGRESS CATALOGING-IN-PUBLICATION DATA

Fountaine, Margaret, 1862-1940.
Butterflies and late loves.

1. Fountaine, Margaret, 1862-1940.   2. Butterflies –
Collection and preservation.   3. Entomologists –
England – Biography. I. Cater, W. F.   II. Title.
QL31.F67A3   1987   910.4'092'4   [B]   87-4919

ISBN 0-88162-307-5

First published in Great Britain 1986
© W. F. Cater 1986

Photoset in Imprint by Ace Filmsetting Ltd., Frome, Somerset
Made and printed in Great Britain by
St Edmundsbury Press, Bury St Edmunds, Suffolk

CONTENTS

ACKNOWLEDGEMENTS

When I began editing the diaries of Margaret Fountaine in 1979 it was my expectation and that of the publishers that we could make one book from her twelve thick manuscript volumes. We were mistaken. Miss Fountaine had written too much, had lived too long and interesting a life, for it all to be contained by one book, even of generous length.

Grasping at the nearest thing to a natural break in her narrative, I finished *Love Among the Butterflies* at the point where she sailed, in 1913, into Brisbane to begin a new life. There were still three volumes of her diaries and many years of her life to be recounted.

When the book was published, editor and publisher were immediately surrounded by readers asking, and sometimes fiercely demanding, to know the end of Margaret's story. Their insistence gives me the pleasure of being able to finish that account, in this new book. I also owe thanks to many people for their help: to the director and staff of the Castle Museum at Norwich, custodians of the diary manuscripts; to various museums and archives where I looked for answers to questions which may, at this distance in time, be unanswerable; to the citizenship authorities in Australia and the United States who helped to clear away a little of the mystery surrounding Khalil Neimy's background; to the *Sunday Times* which originally gave Miss Fountaine her chance of publication; to Brigid Keenan and Elizabeth Grice, former members of the staff of that newspaper with me, the first for helping to fill the silence after the end of the diaries, the second for finding some of the dwindling group who had known Miss Fountaine; to Mrs Grace Veivers of Kuranda, for her recollections of Miss Fountaine in Australia; to Mr Frank Sealy of the Trinidad High Commission in

London; to Dr W. J. Eggeling who remembered the lady with affection. But so, I think, do all of us; those who knew her and those of us less fortunate who met Miss Fountaine only through her diaries. She captured a lot more than butterflies.

ONE

New Readers Start Here

Family and girlhood —— formative first love ; and
money —— travels, men, opera and butterflies —— Fate
in a tarboosh —— love but never marriage ——
adventures together —— the diplomat's proposal ——
rejection and new travels —— the high passes to Tibet

There is no doubt that Septimus Hewson was a cad. He was a feckless, dishonest drunkard, and he broke Margaret Fountaine's heart. There is no doubt, either, that in breaking her heart he helped break the chains of Victorian convention and so enabled her to live the life she loved – a long and, in part at least, a happy life. Without Septimus she might have declined into good works, genteel sketching and bridge parties, and the window she opens for us on most of the earth's surface and on a vanished world would have remained shut.

There is no doubt that he was a cad : he didn't marry Margaret when she thought he was going to. It might be pleaded in his defence that if he had married her, as she so desperately wished, they would both have lived unhappily ever after. The fact remains that by the standards of the year 1889 he behaved unpardonably, twisting and weaving and ducking as Miss Fountaine, a young woman of considerable presence and determination – in later years she was to vanquish Turkish bandits, Austrian Customs officers, hostile tribesmen and drunken Colonials – pursued him with the implacability of true love. Reading the diaries of her earlier years one is desperately sorry for Miss Fontaine, and almost equally sorry for Septimus.

If we owe a great deal to Septimus for setting Margaret to travel the world, we owe no less to Uncle Edward Fountaine, of whose life we know little but whose death brought Margaret an income enough to make her independent. And we owe much to Margaret's mother, dear Mamma, whom Margaret heartily disliked. The diary speaks of a childhood of 'one meaningless punishment after another', of 'the vile way she treated me', of 'the boredom of life with Mamma'. Mamma bored her daughter out of the family circle, out of the county, out of the country, until that young woman found herself, on her first trip abroad, beneath skies more blue and sunshine more hot than England's, and never again really wished to return to the British Isles, with their cold skies . . . and cold men.

It was 1862 when Isobel presented her husband John Fountaine, rector of the tiny parish of South Acre in Norfolk, with their second child and first daughter, whom they were to name Margaret Elizabeth. There were six more children to come: five more daughters and a second son.

The Reverend John Fountaine died in 1878 and on 15 April of that year his widow, along with John junior, Margaret, Rachel, Constance, Evelyn, Geraldine, Arthur and Florence the youngest, moved from South Acre rectory to Eaton Grange, a house in Norwich. Though the rector had been but a country clergyman, both he and his wife were well connected and money does not seem to have been scarce, at least at first. The house was large, and there was a governess, a children's nurse, a cook, at least one maid, and a gardener; possibly other servants. The boys were sent away to school but the girls took their lessons at home. They could enjoy an extensive social life, in the manner of the 1870s, without going outside an extended family circle – there were 22 aunts and uncles, and legions of cousins.

It was a well-off and well-connected County circle; the widow and her family may have been poor relations, but they were well related and only by comparison poor. Years later, Margaret was to remark to her diary how the attitude of snobbish American ladies toward her would change if they knew she was related to half a dozen aristocratic English families. And *Burke's Landed Gentry* and *Burke's Peerage* bear her out.

Margaret began the diaries that were to fill and record so much of her life as a schoolgirl attempt to describe that one day, 15 April 1878, when the family moved into Norwich. She continued to record that one day every year, though the tail of events leading up to each 15 April grew longer and longer until quite soon the diaries were a straightforward record of each whole year, odd only in that they ran from April to April, not from January to December.

One thing that becomes plain early in the diaries is that Margaret was susceptible. At 17 there was Woodrow: '. . . I love the long straight road because it is there we generally see him . . . I love the little church because he has been there.' By 19 there was young Gerald Bignold: dancing with him made her 'madly excited and strangely happy'. The following year young Gerald was displaced by Mr Swindell the curate, 'of quite low origin, risen entirely through his own ability . . . this only made me love and admire him all the more'.

Mr Swindell was, unfortunately, married; but, emulating the heroine of a novel she was reading, Margaret went on sinfully hankering after him – until Mamma took the girls away for a holiday in Cheltenham and Malvern, where Margaret fell in love with one of the bandsmen.

She was 21 when Woodrow, Gerald, Mr Swindell and bandsmen all alike were forgotten for Septimus Hewson, an Irishman of her own age, a cathedral chorister. It was a heartfelt passion, but passion that descended no lower than the heart. Indeed, a year passed before Margaret quite realised what all this passion was about.

She told her diary: 'My eyes were opened . . . I knew why I loved Septimus Hewson. I was not ignorant before. It was only that the idea had never come before me.' Now that she knew, she hated herself for it; but still spent long days sketching in Norwich Cathedral in the hope of seeing Septimus, attended Sunday services there twice a day to hear him sing, contrived to secure a photograph of him, and wrote desperately and endlessly in her diary about him, page after unhappy page.

Mamma, scenting that something was wrong, despatched her to stay with some of the horde of kinsfolk. The group included two eligible young men, but Margaret returned to Norwich still infatuated with Septimus, who plainly wasn't eligible at all.

It is hard, at this distance, to realise the attraction of singers like Septimus to impressionable young women; the nature of their duties and the sacred edifices where those duties were undertaken might be expected to cool the ardour of many admirers. In their own time, however, they appear to have had something of the effect of young secular male singers in our days. Young women – Miss Fountaine was not alone – suffered doubled and redoubled church attendance to gaze upon them and to listen to them; choristers like Septimus drew admiring crowds when they sang at garden parties and other public functions.

At any rate, Margaret and Septimus eventually met and exchanged a few words; she continued to pursue him, to linger where he was bound to pass, to find excuses for conversation, even eventually to write to him unbidden – where was the maidenly Victorian restraint we are told our great-grandmothers had? – to declare: 'I feel so strongly attracted towards you. . . .'

There were misunderstandings, letters smuggled to the post box, replies to be collected, blushing, from the post office. She wrote to Septimus 'in the passion of my despair', asking him to meet her in the cathedral. She waited in vain. 'I had cast aside all the laws of modesty and propriety by writing to him to propose a meeting,' she told her diary, 'and he had stayed away to show me he did not care. Going out after the service it seemed as though voices were calling after me words of bitter scorn.'

She knelt beside her bed that night to say her prayers and 'all self control gave way and I cried such tears as I had never cried before, I who have always had such a contempt for weeping'.

We can, as Voltaire pointed out, bear with remarkable fortitude the pains of others, and this is probably most true of the pains of love. Miss Fountaine's love was absurd – like most of our loves – and her diary does go on about it at an inordinate length. But it is hard not to come close to weeping for that sad young woman (she was 25 now) gradually dropping off to sleep and telling herself with contemptuous despair: 'Now I shall never feel his arms around me and his breath on my cheek. I shall never clasp his children to my breast.'

Margaret was crushed with sorrow but not defeated. After a holiday with yet another relation (during which, sorrow not withstanding, she appears to have flirted mildly with a cousin's personable swain) she was back at the siege of Septimus. One thinks of her as a small, ill-trained but relentless army, defeated and retreating but only to regroup before advancing again. Once more she fell into conversation with Septimus. Lost to all feminine commercial wisdom, she scorned Mamma's matchmaking attempts and a young heir to £30,000: 'To be the wife of any other than Septimus Hewson . . . the thought was madness.'

But Septimus, alas, fell victim to the demon Drink, and the cathedral sent him away 'on account of being such a dreadful unsteady man'; he slipped out of Norwich by night, leaving his debts unpaid. She traced him, sent him a hard-saved five sovereigns. He wrote once from Limerick in his native Ireland, thanking her, but did not write again. What could she do now?

With splendid timing old Uncle Edward Fountaine died and, to everyone's astonishment, left £30,000 to be divided among the six Fountaine sisters and a cousin, giving them each a safe income of £150 a year; perhaps £5000 in today's terms but far more in its purchasing power abroad.

Margaret had no doubts about a suitable use for such providential gifts. Realistically, she set out in her diary her problem and a possible solution. Septimus had taken from her all power to love anyone else. He didn't care for her but she still loved him. 'My only chance now is to attract him by tempting his cupidity and then try to excite his passions . . . he will be much more disposed in my favour as a monied lady.'

With her sister Evelyn as a singularly ineffective chaperon, she sailed for Ireland for a holiday, writing Septimus a careful letter: she would never be really happy without him, but she was now well off and independent and so 'would soon have every reason to forget you. Do you ever come up to Dublin?'

She was, she told her diary, 'not offering myself for sale to a rich husband as many other women had done, but offering as it were to buy the love of the man I love'. One can deplore Miss Fountaine's folly; even today some might chide her unladylike seizing of the initiative. But whose heart could fail to lift as she leads her regrouped army back to the attack? This was the sort of woman about whom young Mr Shaw was to write his plays. Did he, perhaps on some bicycling expedition, fall in with the Norfolk rector's daughter and come away, his mind filled with speculation and endless dialogue? She would have deplored Shaw's political views, but if ever his Life Force were to be seen in relentless action, here it was, as Miss Fountaine restlessly awaited a reply to her letter.

Who could doubt what it would be? Septimus had appeared cool for her sake, because of the difference in their positions, he wrote; but he had been thinking of her. Now she was independent, 'my dear, you must write me a long letter', and he was hers, with love, Septimus.

More letters followed, while she, part fond foolishness, part cool tactician, considered how best to advance her campaign. It went well. 'I will never be properly happy until I have you,' he was writing as she set off for Limerick.

And there, in the moonlight beside the Shannon and in the shadows under the trees, she was at last in his arms and, 'against

13

all the promptings of my nature', exchanging kisses. Septimus gave more than he got but, then, he had perhaps had more practice at this enjoyable and inexpensive form of generosity. It was her first kiss. She was 28. Septimus pledged his true devotion. Admittedly neither then nor in the other days they spent together at Limerick did he ask her to marry him in so many words, not even when he picked the honeysuckle she pressed between the pages of her diary; it is there still. But proposals of marriage were often an understanding that needed no words, Miss Fountaine wrote. In transports of happiness, though with a certain apprehension, she returned to Dublin, to England, to Mamma who was, she recorded, 'very much surprised'.

The lovers exchanged letters almost daily. Mamma wanted to know Septimus's financial position. The trustee of Margaret's inheritance, the millionaire industrialist Sir John Lawes (yet another uncle) was not sympathetic; Miss Fountaine's blood ran cold (her own words) at the thought that he might be able to take her money away from her and then 'my chance of happiness would be gone'. But the kindly Sir John only tutted.

Anxiously she awaited Septimus Hewson's reply to Mamma. It did not come. She waited, firing off letters into the silence. No reply came. Through the end of August Margaret waited, through all September, tormented by fears that he was ill, that he had met with an accident; and tormented by her conscience – it had been a low, degraded scheme to buy happiness. The days shortened and the leaves turned, and every day Mamma asked why there was no letter.

When a letter from Ireland finally came, it was from the aunt with whom Septimus had been lodging in Limerick. Septimus had no idea, said his aunt, that he had entered into any engagement and was in any case totally without means to marry. Not unsympathetically she added: 'He is not in any way worthy of you and I scarcely think him capable of caring much for anyone but himself. We are deeply grieved that such trouble has come to you through him and will pray that you may be helped to forget . . .'

That was that. Half a century later Margaret Fountaine, locking away her diaries for the last time, was to say still that the greatest passion of her life had been for Septimus Hewson.

A year later her real travels had begun; first to Switzerland with her sister Florence, in high summer. 'We had now left the English climate far behind . . . I had never known till now how beautiful was this earth.' They arrived in Berne on a warm night to find the city brilliantly lit and crowded with people costumed for next day's fête, while 'the August moon, red and full, looked down upon the scene. It was like walking in a pantomime.' From that moment, Miss Fountaine was lost to England; she continued to be proud of her native land, but she preferred to love it from a distance.

In Switzerland she began butterfly hunting, discovering in herself a born naturalist, recognising in a moment butterflies she had seen only as a child in pictures, 'when I used to look with covetous eyes at the plates representing the rare Swallowtail or the Camberwell Beauty'.

There was a little flirting, too, before they went on to Dijon, to Paris, and home to England. The following February she was off again, this time with a pair of cousins, to Paris, Nice, Naples (where that buffoon Signor Scafini made advances), Rome, Florence, Milan, Lake Como . . .

Back in England for the winter, she reflected: 'I suppose it is necessary to one's moral digestion to swallow so many degrees of district visiting, Blind Asylum Fridays, Charity Bazaars etc etc, to counteract the delights of roaming over foreign lands with a tolerably well-lined purse.'

Next year it was Milan again, and singing lessons, and a flower – 'the messenger as it were of an unholy passion' – left on the crimson cushion of the opera box after Verdi's *Falstaff*. Butterfly hunting continued, too, between the singing lessons. She was flattered to be asked by her tutor whether Mademoiselle did not desire to sing in the theatre; she had a good and a strong voice and it was not too late to study; she was 21, 22, was she not? Miss Fountaine replied by confessing that she would never see her 25th birthday again; 'true enough to the letter,' she remarked to her diary. She was in fact 31.

She was tempted but came eventually to the conclusion that opera was not for her. She believed that success for a woman in that world depended on what a coarser generation calls the casting-room couch. Anyway, men around the theatre were an unattractive lot; Socialists too, she dared say.

All the same, there was a delicious *frisson* in contemplating the peril into which one might fall, before firmly putting temptation aside and travelling on. Venice, Corsica (where she drank far more than was good for her at a little wayside wineshop and suffered both a hangover and a conscience), Switzerland again, back across the Simplon Pass into Italy, meeting the charming Dr Bruno.

The following year, 1894, she was again in Switzerland, Italy (Was Dr Bruno going to propose marriage? Should she accept him?), the South of France, Lake Como; in 1895, Spain and Sicily – 'Oh, I never spent one dull moment when I was at Palermo'. Nor did she, between the proposals and propositions of the handsome Baron, the advanced-in-years Professor, the handsome young Pancrazio (but 'I could not so lower myself as to allow the son of a hotel proprietor to kiss me'), and the distinguished fellow guest who asked with some surprise had she never *tried* the course he was advocating? And then there was Amenta, the charming boy in whose arms she lay but whose further advances she denied: ah, life in England was never like this.

Miss Fountaine contrived to spend a great deal of time butterfly hunting as well; but all the same, back in England and visiting an aunt in Bournemouth, she found it hard to retail her overseas adventures without considerable modifications.

Next year there was Germany, Austria, Hungary (where the bold Dr Popovich, told that she collected 'les papillons de jour', replied 'je suis un papillon de nuit' and tried to kiss her), Italy again, where she pedalled on her new bicycle from Genoa stage by stage to Venice, and then from Trieste to Fiume ('deadly tired in spite of eight large tumblers of beer') before travelling on to Budapest where the local naturalists were full of those 'little thoughtful attentions so gratifying to the female intelligence'. Back to Milan, home to England, away again to France, Switzerland, Monte Carlo, Athens, Delphi (Marcos the guide was 'well favoured and attracted to me', which she had to admit was an advantage; he didn't like the snakes in the marshes of Mesolonghi but he had to put up with them until Miss Fountaine had captured *P. Ottomanus*) . . .

It was splendid to be free, she wrote in her diary: 'Freedom is the crowning joy of life. Thank God there are few on earth I really care for; I would there were none. I want to see all I can of this beautiful world before I have to leave it, and life is so distressingly short . . . it is the affections that tie us down to one spot . . .'

She returned home, and brought home, too, her sister Rachel from the sanatorium where doctors had now given up all hope

for her recovery; for Rachel at least life was to be distressingly
short. Margaret was soon off again on her travels. It was 1901;
her hunger to see more of the beautiful world, and her ever
more demanding hobby of butterfly collecting, joined in demanding that she go further afield: to Beirut, to Damascus.

It was there in Damascus that her fate was awaiting her, as
unlikely as could be imagined. She hired a dragoman – a guide
and interpreter. He was Khalil Neimy, a fair-haired, grey-eyed
young man of 24. She was 39. Before long he was making passionate declarations – in his strong American accent; he had
lived for two years in the United States – of his love for her. She
despised him and reproached herself for allowing it to continue.
'But how could I help it? The man swore he had no wish on
earth except to make me his wife. I didn't care a damn about
him. But I began to find his untiring devotion and constant
adoration decidedly pleasant . . .'

And so – she never really explains, even to her diary, how it
came about – one glorious summer morning when they had
been hunting the big brown butterflies in and out of the shadows
of the great rocks at Baalbek, she agreed to be his wife.

She allowed him thereafter, she says, some liberties; but not
many, and she regretted a streak of coarseness in him, as when
he would remark, 'I love very much your legs.' She attempted
to raise the tone of his mind. But she had, she told her diary, at
last perhaps found someone who would love her truly and care
for her.

It cannot be said that the course of their love ran smooth.
They travelled the countryside around Damascus, Khalil snatch-
ing discreet kisses whenever he could, Margaret waveringly
responding. Meeting respectable English folk reawakened her
sense of 'the degradation of the course I was following', she
confesses to the diary, 'and so poor Khalil had to sit alone for a
long time before he could persuade me to come and be kissed'.
All the same, lunching demurely with the British Consul and
his wife, she enjoyed a far from respectable thrill recalling how
'only a few hours earlier Khalil had been carrying me across the
room, laying me on my bed, kissing me and leaning over me . . .'
He never leaned too far, however: she 'always denied him one
thing he asked for, so he was very anxious that we should be
married with as little delay as possible'.

There were complications, of which, perhaps, the fact that
he was a member of a different church, the Greek Orthodox,
was the least. More to the point, Khalil was – or so Miss Foun-

taine appears to have believed – a subject of the Turkish Empire. Bad enough in those years, when British consciousness of Imperial destiny was at its height, for Miss Fountaine to lose her British citizenship by marrying a foreigner, any foreigner. Vastly worse for a Christian English gentlewoman to become, by such a marriage, herself a subject of the corrupt, barbaric and internationally despised dying Moslem empire; a subject of that ruler of nameless horrors, Abdul the Damned!

No, after their marriage they would live in the United States, and never set foot again within the Turkish Empire unless they had both become naturalized Americans . . .

There is a mystery here, which we must touch on again later; it is not the only mystery concerning Khalil. Suffice it for the moment that the eminently sensible plan to seek American citizenship is never mentioned again in the diaries, and that each time conventional matrimony seems about to descend upon Miss Fountaine, some new obstacle always arises. Never has marriage so eluded lovers seemingly so desperately sighing for it. Unmarried, however, they continued to travel together: in Syria, Jordan, the Holy Land, before she returned to England and he to his less personal devotion to other travellers.

She awaited his letters with impatience turning to anxiety. When at last a letter with a Turkish stamp arrived, it was not from Khalil but from one of his associates, asking if she knew his whereabouts: Khalil's wife was becoming anxious about him . . .

Miss Fountaine was appropriately shattered: but not, after a short spell of rage and despair, too shattered to draft a discreet inquiry to an Englishwoman in Damascus, asking if Khalil were married, since she planned a long trip and would not take him away from his family, if he had any. No, came the reply; he was not: there would be no obstacle.

Relieved, delighted, Miss Fountaine set off once more for Beirut. In her hotel she waited all day, picturing Khalil's journey there from Damascus; before nightfall he would enfold her in his arms again. A knock came at the door; not Khalil but a missionary acquaintance. 'The man is married,' said the missionary, 'he has been married three or four years and had two children, both dead . . .' She bowed her head beneath her grief; but raised it again when the missionary endeavoured to improve the moment by 'expatiating upon the subject from a religious point of view, at which I became so distressingly blasphemous that he had to beat a hasty retreat'.

Within a year the lovers were together again. It is true that she began by declaring that there must be no return to their old relationship, but not even Margaret could have believed it. Very soon they were hunting butterflies in the wilds of Turkey, and in the intervals of that pursuit, 'ours was a flinty couch, maybe a cavern among the rocks, or some tangled thicket . . . the world would never recognize our ties, so the world must never know; that was all; and children we could never have, even if my advanced years' (she was 41) 'would not anyhow have precluded us from this greatest of human happiness . . .'

That was in 1903. The following year they were in Algeria. Both fell desperately ill with malaria; Khalil's devotion, then, seems to have taken away any remaining doubts she might have had about his love and sincerity, married or not. In 1905 they were in Spain; Miss Fountaine danced with the stationmaster on the platform of a small mountain station 'at the urgent request of the assembled company', and Khalil kissed her every time the train went through a tunnel. Miss Fountaine was, in fact, at this time engaged to be married to the British Vice-Consul in some small Turkish town whom she had met on her travels, another entanglement she seems to have drifted into with surprising helplessness for a woman of so firm a character; but after some vacillation she rejected him out of hand. As she declared in her diary, he was not only twice Khalil's age, but showed every sign of trying to restrict her freedom to travel third-class on trains if she were so minded; and certainly would object to her dancing with stationmasters . . .

In 1906 they were in Corsica, in 1907 cycling together through what is now Yugoslavia. Khalil took off his jacket in the heat, Margaret shed the skirt of her dress and ran about in her cotton petticoat: 'The Turkish peasants were no doubt under the impression that our lack of attire was a national British costume.' Khalil became very ill with pleurisy; when he recovered she decided that they must spend the winter in the gentler climate of South Africa. Through 1908 and 1909 they were travelling and collecting, first in South Africa, then in the raw new colony of Rhodesia ('of course they were all drunk; Englishmen in Rhodesia always are'), and over the border into Mozambique.

The following year Miss Fountaine decided she must visit her younger brother, Arthur, whom she had not seen for more than twenty years. Arthur lived in Virginia, where he had been sent, ostensibly to learn farming but more probably to get him out of England. Mamma made him an allowance, which she took care to preserve by the terms of her will. Arthur was the black sheep of the family. Quite what his offence was, we shall never know. He is rarely mentioned in the earlier diaries. He was certainly given to liquor. Whether he had fallen from grace in any other way we are not told. But Margaret had a long-suffering elder-sisterly affection for him, and was grieved that in those years of banishment 'his poor life had been a wretched and depraved one.'

He was already muddled with drink when he met her off the morning train at Covington, a little town she roundly declared to be 'nothing but a sewer of drunkenness and profligacy'; but there was a young woman, Mollie, who seemed to have some attachment to Arthur and might make something of him. True, Mollie could not help falling in love with the exotic stranger Khalil, but Khalil sternly ordered her to marry Arthur instead, and not long after Margaret and Khalil had whirled on their way (to Havana and Jamaica; the collecting was very good), a letter told them that Arthur and Mollie had indeed married.

Costa Rica in 1911, India in 1912, Ceylon, Sikkim and the high passes of the Tibetan border in 1913; the travels continued. In October of 1913 they sailed for Australia by way of Penang, Singapore, Batavia, Macassar, Thursday Island and Port Moresby. Early in 1914 they sailed into Brisbane. For a time at least their wanderings were to end. A different adventure was waiting for them.

TWO

———— ✦ ————

Ghosts, Madness and Defeat
1914

Australian exile ——— Khalil's chance to be British ———
a 'loose adventurer' ——— looking at farms ; £125 for 70
acres ——— local advice : Don't ——— Khalil becomes 'my
cousin Karl' ——— building a house ——— wattle trees in
golden bloom ——— Khalil returns drunk ——— vile
accusations because of a boy's smile ——— I said I'd
drown myself ——— our ghost-haunted house, what
wickedness once here? ——— whispering in the night ———
and Khalil raving ——— driven out by persecution ———
no wedding and no passport

It was doomed from the start.

The primary reason for the Australian adventure must have
been to enable Khalil to acquire British citizenship. Except for
very brief interludes when emotion overcame her prudence,
Miss Fountaine was determined not to marry until she could do
so without losing her own nationality, or possibly without ac-
quiring Khalil's somewhat indeterminate nationality. So Khalil
must become a naturalized Briton, and for this he must reside in
the British Isles – impossibly cold and uninteresting for Miss
Fountaine, quite apart from the social chill she would inevitably
experience when it became known that she was to marry a
Levantine dragoman – or he must reside in one of the British
Dominions.

There were other reasons for the move : 'to make our fortunes',
as she says; to attempt a less peripatetic life; perhaps even to
give Khalil an opportunity to step out, a little, from his sub-
servient role and to establish himself.

But it was doomed. They were to farm; neither had experience
of farming. They were settling in a new country; neither had
paid even a passing visit to Australia; and if Margaret had
troubled to read anything about her intended homeland in
advance, her diaries make no mention of the fact. Khalil, for all
his fair hair and blue eyes, was plainly a foreigner in then-insular

Australia (even though Margaret took to calling him first Karl and later Charles). Miss Fountaine, very aware of her aristocratic connections, was moving into self-consciously egalitarian rural Australian society. She had capital but was not releasing enough for the venture to make up for their lack of training, though this was perhaps just as well. Above all, she lacked the temperament to deal with the grinding monotony of pioneering. Under the circumstances, she and Khalil managed better than might be expected. It is surprising they lasted so long.

Her dislike of Australia began with the Brisbane Customs; and after only a few days she is already painting herself as that character familiar to Australians: the complaining Briton, the wingeing Pom . . .

We were made to open every single thing, but luckily had nothing upon which they could claim duty, as our funds were running very low indeed. Whatever luck is in store for us in this country, we will not forget how short we were when we first landed. I can't say I was much struck with Brisbane, still less so with its hinterland; a spectacle of bare ugliness which I could not think of settling down in, so we decided, as soon as my next Letter of Credit arrived, to go back up the coast to Northern Queensland, and try our luck there. I booked our passages to Cairns, and we started off to seek our fortunes at last.

I remember when I was quite a small child I told my mother one day: 'Mamma, when I grow up, I mean to be a loose adventurer', and I could not imagine why my mother rather reproved me for the remark, my idea of being a 'loose adventurer' having been to go to Australia and break in wild horses.

And now I was in Australia, the land of my childish dreams, but how different was I finding it. We did not care much for the passengers on this boat, and indeed we had already begun to experience that nearly all the Australians are commonplace in the extreme, especially the women and girls. It was nearly three hours before we could get our big luggage from the hold at Cairns, as there were about 70 horses on board, all of which had to be landed first; so we only just caught the train up to Kuranda, where we fixed ourselves up very comfortably at the Barron Falls Hotel. Mr Hunter, the proprietor, is himself a land agent, and with him we are looking at farms and selections, though at present we have not selected anything. We are 'cousins' here, Khalil and I, and I always call him 'Karl' now.

Yesterday afternoon we went with Mr Hunter to see Mrs Fallon's farm, about two-and-a-half miles from here. But Mrs Fallon was rather screwed, and I did not like the house at all, neither did Karl, so I don't think it will be there that we shall find our much longed for little home. Will it be on that strip of high ground not far from here overlooking the loveliest scene of bush and scrub land, where the cockatoos fly and screech, and build their nests in the hollows of the big forest trees, and the huge *Papilio Ulysses* floats along in the warm sunshine? We went on to visit this spot again this morning, its grassy slopes half covered with young trees, grown up again since it was cleared, and dense jungle which would be of little use to us, except to breed butterflies and be a joy forever in its wild luxuriance. There is no house, and though I can picture what an ideal spot it would be to live in, and what a glorious garden we could have up there, full of all kinds of flowers of the kinds most beloved of butterflies, there would be little or no money in it, and goodness knows how much time *and* money we should have to spend before the house, with its necessary stables and outbuildings, would be completed; one year at least, I think, and that would mean another £250 for our board and residence at the Barron Falls Hotel in the meantime. The property is cheap enough – £125 for 70 acres. The adjoining estate – a much more desirable acquisition – is £250 for 80 acres. But terms are easy in Queensland for buyers are few and most of the land round here appears to be in the market just now, a somewhat suspicious state of affairs.

After lunch I set a few butterflies we had caught this morning, though the one chance of capturing *Papilio Ulysses* had resulted in my shattering what had apparently been a beautiful specimen; he still flies, though in a damaged condition. At 3.30 Karl came over, dressed ready to go with me to see Mr Dodds' collection; the old man gave us a most hearty welcome, refusing to take the usual shilling each for admittance, saying that we were fellow entomologists. He knew me by name, having read my articles in *The Entomologist*, and in the *Transactions of the Entomological Society of London*. When we had talked entomological shop for some considerable time, he began to discuss our settling in this neighbourhood. He did not seem to think the prospects of farming very bright, and took more or less the same pessimistic view that Mr Hunter takes of our intended venture. But as we are not going to sink any capital, only my income as it comes in, which will always be behind us, we certainly can come to no harm, even if Karl does not make the little fortune we both so much hope for, and after all the most important point is that he should become a naturalised British subject out here in Australia, and this is a point we do not deem it advisable to bring before our friends at present.

After dinner we went out into the dark night, for the Easter moon does not rise now till late, and we talked of our farm, and future prospects. Afterwards Karl got sleepy as usual, and with a comprehensive wave of the hand, wished us all good-night and retired. I was just thinking to follow his example when we heard the report of a gun from the back of the hotel. Mr Hunter had shot a snake which had come after the fowls, a big one it was too, over 10 feet in length. We all went out to see it, laid out in front of the hall door. It was, however, a harmless kind – a carpet snake, Mr Hunter called it. Soon after this I slipped away, glad to get back to my bedroom in the cottage; and it was not long before I was inside my mosquito-net and sound asleep.

That entry for the year up to April 1914 was not copied into the bound diaries until June 1921. The next year's record, for the year up to 15 April 1915, is one of the most brief in the diary; without travel there was little to record, and it is a despondent Margaret Fountaine who confesses:

Looking back upon the year that is gone, I feel a vague sense of failure and regret, for it has been a year of disappointment and tears. Maybe I am longing for that old life which was so sweet, for we have laid aside our butterfly nets now, and given up our life of wandering to lead (for, I suppose a few years at least) what is generally spoken of as life in the Australian Bush.

Visibly striving to make the best of a bad job, she goes on:

There is much in this bush life that is greatly to my liking, though we are tied down to some 160 acres of freehold land, mostly standing scrub. We have a house of our own which I had built to suit our own liking on this our own property. We have a garden, in which the stumps of many scrub trees are still flourishing and sending out fresh suckers, though we hope in time to burn or grub out all of them. We have a paddock too, which is one mass of fallen timber lying amongst the grass, and in this paddock we have our own horses so that whenever we care to take a ride through scrub-tracks and up and down slippery creeks we can do so. Of course I love the riding, no matter how rough . . .

Through the cooler months of the Australian winter, Margaret and Khalil had been felling trees on their new property – at Myola, two or three miles from the township of Kuranda – while Bill Moule, a local builder, had been at work on their house. A photograph pasted into the diary is taken through a tangle of fallen timber and high grass; beyond that the timber frame of the house, as yet uncovered, stands high off the ground on tree-trunk piles, in the local style of the time. Trees fringe the picture on all sides. It is a formidable sight for a homesteader, a vista of back-breaking work to be done.

With the outbreak of the First World War, Moule went off for

24

military training (Miss Fountaine records the beflagged trains bringing in boys from the back country for the same purpose). The house was completed by 'old Newman', an irascible character, though not without reason.

Every day Margaret and Khalil would ride over to their new property from the hotel at Kuranda where they were staying. Newman's hammer would echo among the tall trees; Margaret and Khalil would be sawing and chopping at the timber in what was to become their garden and their paddock. Their horses shifted and swished and grazed in the shade; and beneath one of the big mango trees Tatters, a dog whose devotion rewarded Miss Fountaine for finding him starving and adopting him, stood guard over their saddles.

At the end of September 1914, they went to live in their new house where 'day after day the Australian sun burned fiercely on, while in the depths of the scrub the cockatoos were busy building their nests, and often the cry of the laughing jackass* was the only sound to break the silence of those long, hot summer days when the wattle trees were all in blossom by the creek near our house'.

Margaret's financial position, because of the war, was less idyllic. The house now being finished, she owed Newman and Moule £250. She had intended to sell a small investment to raise that amount, and to pay off money owing on the land from her income as it came in, but the war had shut the London Stock Exchange and she could not sell her shares, which would have brought little at such a time anyway. 'I had managed to pay old Newman £100, largely owing to the kindness of our nearest neighbour, Mr F. C. Hodel, a man apparently of ample means and a kindly disposition,' she wrote, 'and he cashed for me more than one cheque. But time went on and all prospect of obtaining the further £150 seemed remote, and Newman was far from pleasant.'

She wrote to Cropper, the accountant in London who looked after her affairs, asking him to raise a loan of £300 and cable it out to her; nothing happened and she toiled on, grubbing out stumps, until she fell ill with fever, becoming too weak to sit up in bed unaided. The doctor from Cairns 'poked me in the right side long enough to convince himself I was not suffering from appendicitis, pronounced my illness to be scrub fever, and said

---

* Better known now by its Aborigine name, Kookaburra; a translation confirmed for me by a citizen of Kuranda who pointed upward and said affectionately: 'There's one on that clothesline now; the buggers are everywhere.'

I should not die but that it would last for ten days, which it did.

'During that illness I dreamed I was driving in a carriage through some Oriental city. Suddenly a woman off the street rushed forward and tried to stop our carriage, calling out "He is my husband!" It seemed in my dream that I should never recover unless I gave up all idea of marrying Khalil, still the legal husband of another woman, and I resolved as soon as I got well I must tell him we must break off our engagement.'

After the fever was gone, news came that the loan had come through; propped up in bed she signed a cheque for the impatient builder. Later, 'with the help of Karl's arm I was able to walk to the verandah; how lovely everything looked, strange and far away, the vivid greenness dazzling me. Later – in December, the height of the summer – though I could scarcely sit up on the saddle, I went for a ride on Broncho. When Broncho began to play up I shook till I had to dismount, but that ride did me good, and I began to recover more quickly. When at last I did get quite well, I felt stronger and better than ever, and the rough riding became more and more to my liking, far more than anything else in this "bush life", to which I have thus voluntarily tied myself.'

April 15, 1915, Miss Fountaine's diary day, in the past so often spent in far travels, saw her journeying only as far as Myola station, with henboxes to be loaded on to the nine o'clock train, bread and meat to collect.

A little siding, with a seat outside a tiny corrugated iron erection for keeping parcels in when it rains, is all Myola can boast of in the way of a station. Karl had to stop the train to load the boxes. Often it does not stop but the guard just chucks out the mail bag *en passant*, and if anyone has a letter to mail for Marreba or anywhere else up the line, it is stuck into a hoop of twisted bamboo and handed up to him as the train whizzes by.

She confesses to feeling sleepy: 'I often feel tired now, for the years are beginning to tell on me.' She was 53; one wonders if boredom rather than age was the trouble. The details of her routine, of horses and dogs and hens, of rides in the bush, fill much of the diary now, along with Miss Fountaine's conventional reflections on the war. Occasionally the focus sharpens; as the young men go off to war, 'the scrub tracks are all overgrown now, for there are no eager boys riding through to keep them open . . . the Veivers are running Mr Hodel's property, for with the departure of his son all his interest in the place seemed to die . . .'

She resented living in safety while her country was at war.

26

But what could I do? Be a despatch rider in France, as some of the French women are, was all I could think of; but then I found out that these women were riding motorbikes, not horses, so I went back to apathy.

At last the drought did break, after four absolutely rainless months; and we sowed all our grass seed, and no longer had the toil of going backwards and forwards to the river for water, which we used to carry up in kerosene oil tins on the backs of our horses. For four months we had had two goats for milk as all the cows had gone dry in the drought, and though the goats did not give us very much they cost practically nothing to keep and were sold for what we paid for them. Charles [*the transition from Khalil, it seems, was now complete*] managed them entirely, and got so fond of them that when we sold them and bought a cow he was quite cut up at parting with them. George Wriede built the new hen-house, and ran up a capital cow-shed where Wynetta is milked every morning. And all these little trifles have gone to make up my life during the past year, so small and insignificant and monotonous.

Worse than monotony was to come.

And then only last evening [14 April] the storm-cloud burst over me. For no reason that I could possibly account for, Charles came back from Cairns the worse for drink and after the train had left we stayed down at the station together for two hours in the moonlight, while he raved and stormed at me, heaping every cruel reproach upon me, accusing me of the vilest acts, which God knows I have never been guilty of – acts of infidelity to him such as never entered my head. I did all in my power to pacify him, till at last I managed to get him on his horse, which I had brought down with me to meet him, and we arrived back at our house, where at last he seemed apparently quite satisfied. But there was a far more awful outburst today. After a long and restless night, I awoke feeling very far from well and with a terrible sense of misery; only to be augmented a thousandfold by Mrs Fowler (a horrid woman, who is now our housekeeper*) coming to the window of my room to inform me that 'Mr Neimy' was gone. He was gone, my poor Charles, under some horrible delusion about me, and all his love for me came back to me, and I felt overwhelmed. Why, in the midst of all his fierce protestations of undying love for me, will he never now believe that I love him too? I thought that maybe he was mad, and would do something very terrible. I had heard him walking about in the dark hours, and muttering to himself many times,

---

* Miss Fountaine was unfortunate in her housekeepers. There were three during the Australian years and she quarrelled with them all. To lose one housekeeper, as Lady Bracknell might have remarked, might be an accident; to lose three looks like carelessness.

and I greatly feared that though I had sobered him into reason last night, the delusion and depression had come back to him, and goodness knows where he had gone, or what he would do.

I mounted Fly and rode to make inquiries by telephone. Charles had been seen at the station very depressed, but he returned later that day on the train, and I was waiting to meet him. He didn't as usual come eagerly out of his first-class compartment, but got down slowly, and when he saw me there was no smile upon his face, and he almost made as though he would turn away. We sat down on the bench, and scarcely were the other people out of ear-shot before Charles again began to rave about my insincerity and unfaithfulness, making the most infamous accusations against me, of which, God knows, I am absolutely innocent. He was quite sober, and still I found myself absolutely powerless to convince him of the truth; till at last I felt I had reached the limit of what I could endure from the jealousy of this man for whose sake I had given up every other tie on earth. I got up and said that I would go down to the river and drown myself. I walked over the railway lines and set off through the long grass, while he followed, saying, 'I don't care whether you drown yourself or not, I will drown myself too. We will die together, for I love you! I love you! I love you!' We stooped down, and got through between the bars of the big white gates, and went on and on, down the red clay hill, which led to the river-bed, two people both maddened to despair. We walked on over the hot sands towards the scrubby trees and bushes which are under water when the river is in flood, and there we both sat down, in utter misery and weariness.

The sun was shining strong and bright, for the long Australian summer is scarcely yet on the wane, and the earth was looking its loveliest, but neither of us knew. Some of the time we were both quiet and silent, and then suddenly his ungovernable passion would rise to torment me, while he vowed I must marry him at once, now, as soon as he could arrange it. But I am born an Englishwoman, and I mean to remain one; I will come under no laws but the laws of my own country, and goodness knows they are bad enough on women, especially married ones. He might kill me, if he liked, but until those naturalisation papers came (the application for which is even now on its way to Melbourne) I will *not* marry him and I did not hesitate to tell him so.

How long we stayed under those scrubby trees I do not know. I had at least partially convinced him that it was a lie he had uttered against me; but then I thought I had convinced him last night, and it seemed nothing but the smile of a boy had put the vile idea back into his mind. We went back to where I had left Fly, standing patiently waiting all this time, and we walked while I led Fly by the bridle.

28

After we returned home, we wandered down together to where the new grass is growing well enough on the upper portion of the land, and I told him plainly that I could not go on living with anyone who believed such things as that about me. And yet the most piteous part of the whole thing is that even his most passionate outbursts of rage are mingled with agonised protestations of devotion. He does love me still, I know, though I often think he will kill me . . .

The following year, the year to 15 April 1917, was no better.

There were some terrible months before me, the secret of which I was never able to unravel. Who were these enemies conspiring against me, inventing the vilest lies about me, to my best and truest friend? And why did he, who must have known me well enough, after all those years of confidential intercourse and true friendship, believe the lies, which, according to himself, were being reported about me on all sides? Oh, the darkness of those days, and the still more horrible darkness of those lonely nights, in our ghost-haunted house at Myola!

Our house was haunted, I had known it from the first night I had slept in it, for though a new structure, goodness knows what deeds of wickedness had been perpetrated on that spot many years ago, when Myola was a little township. The dense tropical growth that had long since obliterated all signs of human habitation till we built our house had not stilled the voices that used to whisper in the night. Maybe that was why we had never had any luck here, till now life had become utterly unbearable to us both. Though I bitterly resented Charles's accusations and we fought and quarrelled day after day, I never seem to have thought what *he* must have suffered. The poor thing was on the verge of insanity, though it took me some time to realise it, till one night I heard him singing and raving in his room, and when I went to him I found a man no longer in his right mind. I crouched down on the floor beside him, and struggled to keep him quiet, while I strove vainly to calm and pacify his troubled mind.

Then, one morning I heard Charles's step coming up the stairs, light and joyous, and his voice full of elation. He had just returned from riding up to Mrs Turnbull's for the mail, and he burst into my room with such delight upon his face as had not been there for many a long day. His naturalisation papers had come from Melbourne! He was now at last a naturalised British subject – and there was no obstacle now between us and our immediate marriage. He had told me some little time ago that he had heard his wife was dead, though how much truth there was in this statement I am not prepared to say.

It was the 13th of June, the very day, fifteen years ago, that Khalil and I had sat in the evening amongst the ruins of Baalbek and watched the swallows flying, and I had first promised I would be his wife. He had got the wedding ring all in readiness, but I knew the superstition he always had about the month of June would make all idea of an immediate marriage out of the question, and I soon began to perceive that he did not now seem nearly so anxious about getting married. I was making over the Myola property entirely to him, which, now being a British subject, he was able to hold in his own name. When we left the lawyer's in Cairns Charles told me he preferred to wait till we had got to Brisbane before we married. We went to see Mr Dawson (the clergyman) to arrange the matter with him.

Now I shiver as I look upon the wreck of our lost dreams. We had given up housekeeping and used to take our meals as paying guests up at Mr Rob Veivers', but the nights were horrible and gradually I began to be aware that this man I had loved so was bordering on insanity, that my life was scarcely safe along with him in this lonely spot. Night after night I would lie awake and hear him wandering about the house, opening and shutting doors and acting like one demented. Then he told me he had had a horrible dream of his wife and his father forbidding him to marry. It was evident that this dream had taken a fearful hold upon his already disordered mind, and we both agreed after this that it was better to give up all thoughts of marriage. I was not surprised, for I never did believe this woman was really dead, and I said I fully expected now that we had both come to this conclusion that things would go better with us.

> They did not. Miss Fountaine and Khalil moved out of their house; it was robbed in their absence. Even her little engagement ring had gone. They rode to see friends, and returned to find their henhouse emptied. Animals died from the Queensland plague of scrub-ticks; others were mysteriously driven away in the night. Miss Fountaine received an anonymous letter repeating the accusations that Khalil had made. At last they sold their stock and their furniture, let their house, and fled. Her last action in Queensland, on the day they left, was to vote in a referendum on conscription.

Half-sulky, half-defiant youths were to be seen loafing about on street corners that day in Cairns, fully conscious that the great stake related to themselves . . . I took good care to put 'yes' on my paper, hoping much to see those cowards and slackers, hiding away behind the scrub bushes, dug out and forced to the Front. But alas, the referendum turned down conscription for Australia . . .

They would have left Australia together, but Khalil could not get a passport until he had been a naturalised citizen for at least a year. They agreed it might be better for them to part for a while, Miss Fountaine torn between anxiety at leaving a sick man by himself and the strain of living with his continued suspicions. 'I *would* have married him, I was still true to him but powerless to remove that cloud from his mind which kept forcing him to think I was not.' They stayed for a while in Sydney, going to theatres and 'picture shows'; he tried hard to get a job but without success – experience as a failed horsebreeder and Levantine dragoman were not helpful qualifications in that city. She set him up in a small dry-goods store, and Khalil, still declaring that they must marry some day, arranged to meet her again at her brother's house in Virginia when he should get his passport. A weeping Miss Fountaine caught the San Francisco boat from Sydney docks, wearing a new engagement ring of three sapphires bought for her by Khalil to replace the one stolen at Myola. The Australian adventure was over.

THREE

Long Ago, Far Away

*The puzzle —— the remotest of chances —— climbing
the range —— a magic childhood landscape —— Myola
station —— Mrs Veivers looks back —— when Margaret
rode in from the bush —— such a dear —— a haunted
house? Oh, no —— farmer and American citizen ——
more to Khalil than meets the eye*

That disastrous Australian adventure is puzzling. Australians are perhaps of all people the least likely to ruin the lives of strangers by backbiting, anonymous letters, blows in the dark. If Australians bother to disapprove, they do so unmistakably and face to face. And the Australian landscape can be one of magic or of mystery, but it is not one where ghosts might whisper in the silence of a night-time house. How much of the hostility Margaret felt round her was illusion?

On a more practical level, why did Margaret and Khalil choose, from that whole vast continent, to settle in an area where scrub-ticks might kill half the horses they planned to breed? If they had not read of the dangers before they arrived, they had warning enough from kindly people on the spot: Mr Dodds the butterfly collector, Mr Hunter the hotelier. Why go ahead, and why there?

In 1985 I had an opportunity to return again to Australia, a country I like, and to visit the area where Margaret and Khalil had begun with great hopes. It was seventy years, a fair lifetime, since those ill-assorted lovers mounted their horses and rode, defeated, down from Myola; no chance of finding anyone who personally remembered them, I thought. At best I might find second-hand recollections about the strange couple who had lived there once.

I could look for whatever seventy years of storm and white

ants had left of their house; and at least I might see their view over the wattle trees in golden bloom, find where Margaret had watched the butterflies sailing over their newly cleared paddock, listen as she did to the calls of strange birds sounding over the land beside the Barron River.

Inquiries in Brisbane and Cairns turned up no record of her purchase of land, or of the building of their house at Myola; there was no record in local directories of their stay. It was even hard to find Myola on a map; it is a district rather than a distinct community.

The great wooden quay alongside the Brisbane River where Margaret and Khalil landed is still there – a car park now. The massive Victorian shipping office beside it, at whose long mahogany counter they would have booked their coastal trip to Cairns, no longer transacts much shipping business. The Customs officers whose attention worried the slim-pursed Miss Fountaine are more in evidence today at Brisbane's international airport than at the riverside. Brisbane is no longer an overgrown cow-town, but a modern city with tall office blocks beginning to crowd in toward the Botanic Gardens where she would have walked. And let it be said, if the hinterland of Brisbane were really as bleak and ugly as Miss Fountaine declared it to be in 1914, then the Queenslanders must have made some massive improvements in the last seventy years.

Visitors going on to Cairns, more than a thousand miles to the north, no longer wait for a coastal steamer. They fly, or drive, or if they seek a slightly old-fashioned pleasure they take the train, two nights and a day swaying over the narrow-gauge track in the Sunliner Express, a name more true in its first half than in its second.

Cairns is a thriving town now, a holiday place with smart shops, good hotels, excellent restaurants; a town of broad streets running down to a palm-lined shore where squadrons of pelicans wheel and swoop over the sea or stalk the water's edge.

I sat in my hotel room with the local telephone book and a list of names, taken from Miss Fountaine's diaries, of Kuranda and Myola people. It would be the remotest of chances to find some survivor who in old age remembered them, but I could try. I picked the name Veivers first out of the book. I was looking, I explained, for someone who might be connected with the Rob Veivers who had a bullock team around Kuranda about 1914. At my second call the voice at the other end of the line said: 'Rob Veivers was my grandfather. You'd better talk to my mother.' That lady, Mrs Grace Veivers, was in her eighties, but she was out that day, visiting; she would be home next day.

The following morning I drove northward on the fast road that leads past the airport and between field after field of sugar cane, before turning inland to climb the range toward the Atherton Tablelands. That road, the Kennedy Highway, touches here and there the spectacular route of a little railway which still runs, as it did in Miss Fountaine's day, the fifteen or so miles from Cairns to Kuranda, in and out of the great gorge of the Barron River and on along the plateau. These days the train ride up the gorge to Kuranda is a tourist trip, but Margaret and Khalil would recognize the old but handsomely restored coaches. The small station at Kuranda looks down to the Barron River; a brass plaque records the names of some of the young men they saw ride away to war and who did not return.

Here, in the railway line winding up that river-green gorge, was one explanation of how Margaret and Khalil had come to, had even heard of, Kuranda; a small town now but in 1914 a community of only a few score souls. For at the turn of the century, too, Kuranda was a place for tourists and holiday-makers. Wealthier citizens of Sydney and Melbourne, even people from as far west as Adelaide, would take ship to Cairns; a sea voyage to give them relief from climate, or boredom. The climax would be a trip on the railway winding its way up that astonishing gorge of the Barron River, to a pleasant hotel (in a land where hotels were not always pleasant) in a lush landscape very different from the parched flat miles of a sheep station or the tamed tram-clanging suburbs already spreading round the big cities. Shipping lines advertised such holidays; visitors told their friends. In some such way Miss Fountaine, too, must have heard of Kuranda. She may have learned that land in North Queensland was cheap. She had not, unfortunately, discovered the problems of clearing land or raising stock there.

No matter: Cairns lies 500 miles into the tropics, whose heat and clamorous life always delighted her, as though she were still trying to thaw out from her early years in Norwich, where the chilling east winds blow in across the flat land from the North Sea. But there could have been more to it than that.

Kuranda is already high enough to be clear of most of the damp heat of the coastal plain. It doesn't look like a conventional picture of the tropics. What it does look like is a strangely English landscape, a magic landscape, a countryside of childhood. Turning off the main road beyond Kuranda along a winding lane signposted Myola, with the Great Dividing Range far on the horizon, it resembles that part of England where the western horizon fades into a distant skyline of Welsh mountains.

Here are the small wooded valleys which give an English landscape a human scale; there the meadows scattered with clumps of trees like an English parkland, dotted with cattle grazing or just standing, tails swishing, in the shallows of a small stream from which, as the car passes, a long-legged bird rises, grey wings flapping. There is something familiar yet oddly different about it. The occasional fern by the roadside is familiar enough – except that it towers, seven feet tall, above the head of a passer-by; towers as smaller English ferns do above a child's head. Leaving aside the occasional palm, the trees are not, at a glance, particularly exotic; but they are strange enough to give the feeling of unfamiliarity, of everything seen for the first time, which is the joy and the terror of childhood.

It had rained as I drove up to Kuranda, and there was still a glint of rain in the sunshine – a gentle sunshine by local standards, since this was the Australian autumnal month of May. The woodlands had the heavy greenness of English high summer; an English traveller, again, might expect to turn a corner of the winding lane and come upon a village with towered church and cricketers on a village green, and hear the clout of bat on ball, or even, perhaps, the click of croquet balls on a Victorian vicarage lawn. There is a shock of disorientation in rounding a bend and finding a tin-roofed bungalow with orange trees growing in the garden . . .

Myola station is still there, where the single track splits to form a small siding. Across the line, a small, white-painted shelter still stands, to protect parcels or the odd passenger from the rain. The train to be heard far off, beating up the climbing track, is diesel-engined now, not steam; it passes and the silence closes round again. From the one platform one can look down to the Barron River; the slope is too thick with bushes and tall grass now for any distraught lovers to stumble down to the water, but at this season there are stretches of sand and clay along the margins of the sunken river as there had been when Margaret and Khalil argued tormentedly there. I turned the car and drove back into Kuranda.

Kuranda is a pleasant place, with a prosperous air about it; a commuter township, with sections of good new houses where a signposted trail leads towards an observation point from which visitors can look down on the Barron Falls. The Falls are a shadow of their old selves, say Kuranda people, since the dam was built further upstream; but the town has enough charm to have become, a few years ago, a centre for hippies and other idlers, not to speak of artists; these dissolute characters are not visible to the casual eye.

Mrs Grace Veivers' white and blue wooden bungalow stands in one of the older tree-lined streets, opposite the 'old church' – 'old' here meaning old enough to have a colonial-style veranda.

Mrs Veivers, well into her eighties, white-haired but brisk, lively, very alert, said she had come to the town in 1916; very young, and soon to be married to Kenny Veivers.

The Kenny Veivers who was to move into Miss Fountaine's house at Myola when Margaret and Khalil moved away? Yes, indeed, said Mrs Veivers; and a lovely house it was too.

Was it Kenny's father who had a bullock team, as Miss Fountaine described it – a team of some two dozen bullocks? Why, yes – though Mrs Veivers remembered, with precision, that the teams usually had 22 bullocks, not 24, as they brought in the big logs to the railway station. The Veivers men were all teamsters, one time or another. Her father-in-law had a team; her husband had a team; Bert Veivers, her brother-in-law, had a team; Jack and George both had teams . . . There were no proper roads around Kuranda then.

She remembered Miss Fountaine and her cousin Mr Neimy – a Frenchman, wasn't he? – riding through the bush, of a Saturday or Sunday, to have dinner at the Barron Falls Hotel – Hunter's, it used to be called, after Mr Hunter who ran it. A beautiful hotel, in those days, with six or seven girls working there, and a Chinese gardener; a hotel with a roof garden . . . though that got blown away in the cyclone of 1918, or was it 1920? Miss Fountaine was a very fashionable sort of woman – always loved pretty hats, hats with flowers on them. Mrs Veivers remembered her clearly, riding in, wearing a hat with big roses on it.

What, even riding a lively pony along narrow, overgrown bush tracks? How could she keep a hat on doing that? Her hat, Mrs Veivers explained, would be tied on with a veil; that was what they did in those days. And she sketched a gesture over her own white hair, as of someone twisting and tying a veil.

And the musical evenings at the hotel Miss Fountaine mentioned? Yes, they did indeed have musical evenings, and everybody could go, it didn't matter who, not only guests at the hotel – and at Easter or holiday times the hotel would be full, you couldn't get a bed. They'd give you a cup of tea and a biscuit or something, and you could join in. ('If you go to an hotel now,' Mrs Veivers remarked with some scorn, 'all they'll do is try to sell you a stubby.' A stubby is the standard small Australian bottle of beer.)

We went over some of the names in the diary: pretty young Mrs Franklin – they were two of a kind, Mrs Franklin and Miss Fountaine, Mrs Veivers remembered. The Franklins used to give parties, and in those days if anybody held a party everybody would go; in those days everybody was everybody's friend. There were so few people in Kuranda then. Miss Wreide; yes, she was the stationmaster's daughter at Kuranda, who used to play the piano, and she was great for the church and joined in all sorts of concerts; she would have been a great friend to Miss Fountaine. And the Hodels; well, they lived up at Myola, the whole family, and that would have been a bit of social life for Miss Fountaine and Mr Neimy. Mr Hodel, the father, he was a nice old boy; did a bit of farming, a bit of dealing, a bit of everything, before he sold the place to Rob Veivers who went in for dairying there, and moved into Cairns.

What was Mrs Veivers' impression of Miss Fountaine? I asked.

'She was a dear, a nice person, helped any way she could. A dear – I would have said a dear old soul, then; I was very young. But there was always a bit of humour about her and Mr Neimy too.'

And their house, into which Grace and her husband afterwards moved?

'They had a lovely home; it had two big rooms, a bedroom and a sitting room, with a veranda all round, and then the corners of the side veranda were closed off and made into two rooms, and the back veranda into a dining-room, with the kitchen extended off, and a servants' room . . . a nice house.'

Where was the house? I asked.

Mrs Veivers and her son, who had joined us, considered; it had been a long time ago. And the house wasn't there any more. 'They took it away. Sold it – took it down to the train and down into Cairns, and had it rebuilt near the railway there, near the Grand Hotel.'

I thanked Mrs Veivers and got up to go. But there was a last question. She had lived in that house at Myola. Miss Fountaine had felt unhappy there; thought there was something strange about the place. Out with it. 'She thought the house was haunted. Did you ever feel anything like that?'

Oh, no, said Mrs Veivers, nothing like that. It was a very nice house. She was happy there, and had been sorry to leave when, with Kenny, she had moved away. If I wanted to know any more, I should talk to Johnny Caine down at the motel; his family had the Myola house after she and Kenny moved out.

Mr Caine was deaf and moved with difficulty, but he knew about the house; not personally, I must understand, because he was born only in the year the house was blown down.

Blown down? Yes – in the great cyclone, the cyclone of 1920, but he had heard all about it. He looked at the picture from Miss Fountaine's diary; yes, that's the house it would have been, the sort they were building around the district about 1914. Being all timber, it could have been rebuilt after being blown down, and that's what happened, it was rebuilt in Draper Street, Cairns.

Before I drove back into Cairns, I turned the car toward Myola once more. The lane crosses a bridge above a creek, the stream down which Miss Fountaine once rode, the creek that bordered her land. There's a new house on the land today. A young couple live in the house; they knew nothing of their predecessors who helped to fence and clear this small piece of their country. From the house there's a view of field and trees dipping into a small valley, with here and there a tree standing out from the greenery with its golden blossom. Nearby, above a gateway, a sign says 'Monara Stud'; so someone, after they had gone, fulfilled Miss Fountaine's dream of breeding horses there.

Back in Cairns I tried to find the house Miss Fountaine had built, but without success. Town development must have destroyed it more successfully than the great cyclone of 1920. But the cyclone, the *Götterdämmerung* ending, the warning sky, the rising wind, the house creaking, the gusts pressing harder, the trees thrashing and bending, the storm-note rising to a howl, a shriek, and then the sudden crash as a window shatters, a sheet of tin from the roof whirls away, a snapping and rending sound as a wall goes down . . . it would have appealed to Miss Fountaine's romantic imagination more. A pity she missed it.

These reflections aside, my conversation with Mrs Veivers had confirmed my own prejudice: the hostility for which Miss Fountaine blamed her neighbours seemed to be unlikely. True, Mrs Veivers was then a young girl soon to be married, who might be expected to see the world in rosier colours than Miss Fountaine in her fifties, beset by money and other troubles, struggling against her own ignorance and incapacity to set up a farm, and living for part of the time at least in the terrifying company of a man filled with paranoid suspicion of her. Had Khalil's mood affected her own? Yet Grace Veivers' unprompted recollection of Margaret and Khalil was of two friendly, pleasant people, by no means isolated, living in a small rural community where 'everybody was everybody's friend'; a couple with a bit of humour about them. We shall never know for sure whose picture of Myola is the true one: the friendly people, the happy house, or the hostile people, the hostile, haunted house.

Another puzzle arises from the Australian visit. It will be recalled that while at Myola Khalil received with great joy his naturalization papers making him a British subject and – briefly – bringing again to Miss Fountaine hopes of marriage. I wrote to the Australian Department of Immigration and Ethnic Affairs in Canberra, asking if any further information were available. In reply, with a helpfulness rare among officials, the Department sent me photocopies of all the papers 'concerning the grant of naturalization to Mr Charles Neimy in June 1916 when he was residing at Myola via Cairns. The effect of that naturalization was to make Mr Neimy a British subject . . .'

The first of these papers is an application form dated 11 April 1916 by 'Charles Neimy, farmer'; it declares he is by birth 'a Greek-French subject', after which a different hand has written '(USA)'. It says he arrived in Australia on 4 March 1914, aboard the SS *Van Houtmann*, and had resided 'at Kuranda six months, at Myola near Kuranda one year and six months'. He is, the form declares, unmarried and has no children. Paragraph 9 has been altered from 'I am not a naturalised citizen of any other country' to read 'I am a naturalised citizen of the United States of America'. With it is a statutory declaration: 'My name is Charles Neimy. I was born on the 15th day of July in the year 1877 at Cairo, Egypt. My occupation is that of a farmer'. There is a copy of the oath of allegiance: 'I Charles Neimy do swear that I will be faithful and bear true allegiance to His Majesty King George V, his heirs and successors according to law. SO HELP ME GOD!'

Another certificate is signed by Fred Hodel, a Justice of the Peace – their neighbour at Myola, now living in Cairns – declaring that Charles Neimy is known to him and is a person of good repute. And there is a letter from Mr Hodel, on notepaper with a printed heading 'Wydena Estate, Myola' crossed out and altered to 'Cairns', in which Mr Hodel writes, in support of Khalil's application, that he has known Charles Neimy for a few days over two years; 'he was my neighbour at Myola for about 18 months. I have not the slightest hesitancy in recommending his naturalisation; and hence take this opportunity of covering his application with this more personal recommendation than the ordinary formal certificate which I have given.'

From the handwriting, all these forms appear to have been filled in by Mr Hodel; the paragraphs are initialled 'C.N.' in Khalil's unpractised hand and signed by him, somewhat shakily, Charles Neimy. Here, in Mr Hodel at least, Khalil and Margaret appear to have had one kindly and helpful friend.

On 20 April 1916, the Secretary to the Department of External Affairs writes to Mr Charles Neimy: '. . . I have the honour to request that you will inform me of the nationality and birthplace of your father and mother and paternal grandfather. I have the honour to be. . . .' on 5 May, Khalil replies: 'I am pleased to inform you that my parents were Greek. My father was born in Athens and to the best of my belief my paternal grandfather and my mother were also born in the same place. I have the honour to be, Yours Faithfully, Charles Neimy'.

On 17 May the Secretary observes that Mr Neimy is a naturalized citizen of the United States: 'please state the date when and the name of the County and State in which you were naturalized', and has the honour to be 'Sir, your obedient servant'.

Twelve days later Khalil replies: '. . . I left the United States 19 years ago, and I had lived at Oshkosh, State of Wisconsin. It was in the year 1895 that I received the naturalization papers from Madison (the Capital) but I am sorry I cannot give you the exact date as the papers were subsequently lost, together with other of my belongings, in a storm at sea.' Khalil, too, in the politeness of the time, has the honour to be his faithfully.

Finally, on 12 June, the day when Margaret heard his light step and saw with delight that he was again a happy man, there is a formal receipt signed by Charles Neimy that he has received Certificate of Naturalization No. 23524, to which is appended a letter in Khalil's execrable handwriting, but possibly not entirely in his own words: 'Dear Sir, I have with much pleasure received this morning my certificate of naturalization and I would like to say how much I esteem it an honour to have become a subject of the greatest Empire in the world. Yours faithfully, Charles Neimy.'

But wait a moment . . . The earlier diaries record Khalil telling Margaret of his youthful visit to the United States, of his stay in Chicago, even of his slips from virtue when exposed to the temptations of that great country; they do not record the fact that he was, at the time of their meeting, a naturalized American citizen.

Did Khalil tell her? For if he were an American citizen,* it would have removed the obstacle to their marriage, a marriage for which he had been ardently pressing, at least during the early years. If he were not, what advantage would it be to him to claim American citizenship when he was seeking British nationality in Australia? He was born in Egypt – part of the Turkish Empire. He lived in Damascus, part of the Turkish Empire. His parents were Greek born, but somehow he claimed to be a Greek-French subject, an amalgamation neither country is likely to have recognized.

As in his capture of Miss Fountaine's heart, back at the turn of the century, there was a lot more to Khalil Neimy than meets the eye. But what it was, we shall never know.

---

* The US authorities now confirm that Khalil Neimy did acquire American citizenship.

FOUR

California, Here I Come
1917

*Alone to America —— the little bright gardens of Los
Angeles —— bandages and butterflies —— crossing the
desert —— Arizona, Texas, New Orleans; a spinster
aunt —— the hot blood of the Fountaines —— butterfly
hunting around Hollywood, where people are so friendly
—— Is Mr Piazza too friendly? —— hiring a horse at
Yuma —— Alas, dispatch riders ride motorcycles in
this war*

Miss Fountaine, meanwhile, had crossed the Pacific and made
her landfall on the West Coast. Her discovery of America was
about to begin.

It wasn't long before I began to feel the fascination of Los Angeles with
its sunshine and its flowers in the little gardens with their fresh, irrigated,
grassy lawns, and scarlet geranium hedges – if indeed there be any hedge at
all, for in the suburbs of this lovely city the gardens are just left open and
unenclosed, for everybody to enjoy, as though the people were so filled with
the genial atmosphere that it becomes a common law among them to let
others share in all the beauty; while the golden Californian poppies grow on
every bit of waste ground.

I used to go to the French Red Cross and spend the long days working
with the rest of the ladies, mostly folding bandages to staunch wounds which,
but for the amazing insanity of men, need never have been inflicted. It was a
relief to have found something to do, however little, to help. But towards the
end of my time in Los Angeles I broke away. The sun was shining, the mountains clothed in flowers, while several most interesting species of butterflies
were appearing now that spring was coming to Southern California. First I
tried to give up three days in the week to the Red Cross, and then I let it go
almost altogether and went back to my old free life, wandering over the
mountains and down in the canyons with my butterfly net, just as years ago
I used to wander alone over the mountains of Central and Southern Europe.

Every mail now brought me letters from Charles, who had taken another shop in Kent Street in the heart of the city where he appears to be getting on fairly well.

I awoke this morning, April 15, 1917, in the bunk of a Pullman Sleeper somewhere in Arizona, and soon realised that the dry heat of the Californian desert – through which we passed yesterday in a dense atmosphere of white sand driven before a scorching wind – had given place to the keen chilliness of dawn, and my legs were aching with the cold. By and by I reached up to the empty berth above mine, pulled down a big rug, and then got warm and fell asleep again, and dreamt that I was with Charles, and that he kissed me, but just as I was thinking we would be together again for this day I woke up. The other ladies were all on the move, so I faced the inconvenience of dressing, half in my bunk, behind the curtains, and then carrying off with me all the necessary accessories to a lady's toilet, to the women's dressing-room at one end of the car, a similar arrangement for men being located at the other end. When I had dressed, 'last call for breakfast' was announced – the head waiter told me the time was now 10.15, and on looking at my watch (the lovely little silver watch that Charles gave me on my last birthday) I found it was only just after 8: so the clock has moved already rather more than two hours in advance of Californian time.

We had arrived at El Paso, in Texas. A less attractive looking place I have rarely seen: arid, with reddish dust driven before the wind. No wonder the poor animals looked starved and miserable; what they find to live upon at all I can't think. My caterpillars, five *Papilio Tolicaon*, were doing well, all now in their last skin but one, so I am afraid the next change will take place before I can get to Virginia, in which case I shall miss being able to draw one of them in their earlier stages.

A lot of soldiers got on to the train at El Paso, nearly all very young, and it was good to see the poor fellows, with their tired, dusty feet upon the cushions, some sound asleep. Texas did seem a parched and desolate country, very thinly populated; the mountains too were bare, suggestive of a land without water, while nothing could look more hopelessly rainless than the wide, unbroken blue of the Texas sky. It was very hot and stuffy in the Pullman car; how the Americans stand such a suffocating atmosphere I can't think. As I travelled, I was picturing to myself Arthur and Mollie's boy, Lee Warner, who has just passed his fifth birthday; their other child, little Charles Melville, will not be a year old till May 19. Another boy in between these two died after living but one week, and I fear Charles Melville is anything but strong. I am looking forward to seeing these two baby Fountaines born in America. I feel quite like the old-fashioned spinster aunt, though possibly

43

before the end of this year I may have relinquished that role forever, if Charles turns up, and our marriage takes place.

I spent some time in the Observation Car, and when I left was glad to find my berth had been fixed up for the night, so I at once faced the inconvenience of getting decently into bed, a performance by no means facilitated by the fact that the berth opposite mine was now occupied by a young man, that above me being also taken by one of the opposite sex. My window was still open, and I lay awake some time, and thanked God that at least the fearful conditions upon which this day last year had closed were over, and North Queensland, the most unfortunate venture in my life, had become a thing of the past.

After a short stay in New Orleans I made my way north, till one Monday morning I was up at the Hot Springs and there was Arthur at the depot to meet me. He had brought his old horse and buggy to take me back to Ashford where they are now living; a very emaciated horse, old Sue, now aged some twenty-five years, attached to a rather shabby looking buggy. Presently we came to a halt in front of a house on the high road, and Mollie came out to meet us. Lee, the small boy, who clung close to his mother, was a beautiful child, with the most lovely blue eyes though at the same time bearing a strong likeness to his little Papa. The baby, Charles Melville, was quite unlike Lee, with brown eyes, and a most absurd resemblance to his Aunt Geraldine. I soon loved both those children dearly, and I liked Mollie very much – the cleverest thing Arthur did was to marry her, and she has brought two beautiful children into the family.

I bought a horse, rather against the wishes of Arthur, who seemed to think I ought to be quite satisfied to remain in their house, a muchly paying guest without any particular occupation, and sometimes Mollie would trick out old Sue, and we would go for a ride together, maybe down to the Natural Wells to see her people. At other times I would go for long rides, returning often with a box full of butterflies. Then Lee would look on with an all-absorbing interest at me setting my specimens, albeit the difficulties from a restless boy leaning on my right arm were not altogether desirable. The fact that he really was rather a naughty little boy made him all the more captivating, for, as I used to tell Arthur, he had all the wild blood of the Fountaines in his veins, and though he did chase Mollie's chickens unmercifully, had Arthur forgotten how *we* used to hunt the hens, on Sunday afternoons at South Acre? And if the child did throw rocks at old Sue when she was turned out to grass, we used to chase my father's horses in the meadow at South Acre. But Arthur would not view it in that light, and though he idolized his son, chastisement would have to follow, though I would implore Mollie not to whip the child. I knew from my own experience the injurious influence it cannot fail to have on the character. This boy brought back to me so many thoughts of my own childhood when punishment would be dealt out un- sparingly – and how little good and how much harm it did. I would hide somewhere and cry where none could see my tears, and then defiance never failed to follow. Now I could see in Lee the same childish troubles, sorrow not unmingled with fear, and that same defiance.

I wonder why that curious sense of being hunted and having to hide, to get away from some pursuing evil, has been with me all my life. I remember when I was a child my mother used to say: 'Margaret always has such a hunted look.' Maybe in former life I was some wild animal whose one desire was to escape from its pursuers. How often do I not dream that I am being pursued, hunted down and fleeing away, facing any danger rather than be overtaken by this unseen implacable foe.

The whole state of Virginia having gone dry, Mollie told me that Arthur really had entirely given up drinking now, but the poor thing was himself a wreck, bodily and mentally. I spent four months with Arthur and Mollie; while I was there Arthur had a bad attack of influenza; I had it too, only *I* made up my mind I would saddle my horse and ride to the top of the nearest mountain and fight it out alone up there, which I did, while Arthur went to bed and was really very ill. After that, Mollie fell sick from the strain of nursing Arthur – he would never allow anyone to do a thing for him except Mollie, and poor Mollie, with all the washing, the cooking and the baby to look after, had a very trying time. I suppose I am rather a useless person in the house, but there was nothing for me but to ride away out into that lovely country, which I explored for miles round.

Before I left I allowed Arthur and Mollie to know that I was engaged to Charles, who, because of the war was unable to procure a passport to come to this country, a terrible disappointment to us both. Arthur was quite vexed at my leaving them, not from any desire to see more of me, but regretting the source of income I had been to them, though indeed my own was so materially diminished from the immense income tax I was having to pay that I was quite unable any longer to help Arthur in the payment of his, and I told him so. I went to New York where I stayed at the Martha Washington Hotel, busy getting a big trunk packed with all my store boxes full of butter- flies to send back to England. I could not go myself, even had I wished to, for the submarine warfare was at its height and besides there is a law now forbidding all women to enter either England or France unless on military business. I visited Niagara Falls – nothing like so beautiful or so grand as the Victoria Falls in Central Africa – and spent one day and one night in

Chicago trying to find Charles's uncle in Canal Street, but could find no clue concerning any people by the name of Neimy. Then came the long three days and nights' journey 'out West', through luscious meadow-lands, across huge rivers, past big cities and finally over the deserts of New Mexico and Arizona, till once more I found myself back in lovely California. I was now practically an exile from my own country, and though I grieved not at all for this I shall never forget how this splendid land has given me a home.

Mid-winter in Hollywood, where I had taken up my abode, is like a prolonged and lovely spring. The gardens are one blaze of flowers, and though the continuous drought made the hills bare, the irrigation system is so perfect, like most things in this highly favoured land, that the lack of rain is scarcely felt. Through a somewhat back-stair acquaintance I had managed to obtain with Dr John Adams Comstock (Curator of the South West Museum) I was brought into contact with several local entomologists, some of whom, such as Mr Piazza and Mr H. M. Simms, I got to know quite well. Mr Simms was a young Englishman twice invalided home from the trenches, and here to regain his health. Mr Piazza was half Italian, and though quite English in his manners was not the least like an Englishman in his appearance; in fact typical of a certain type of man who would be described by a certain type of woman as 'very foreign-looking'; a middle-aged man with pleasing manners and any amount of keenness for entomology. I found, almost without think-ing of it, that I was getting on very friendly terms with this man, and it began to strike me that Charles might object to my being so constantly seen with him poking around the Hollywood gardens after larvae (they only occurred on a cultivated Cassia which always grew in private grounds, but the people were just as genial as their climate, so we never failed to obtain permission to search their trees). I also began to suspect that Mr Piazza, a man of any-thing but ample means, was contemplating that the addition of a few extra hundreds per annum to his rather slender income might not be an altogether undesirable arrangement.

So I decided I would go to Arizona, having been told that Yuma was the best locality for *Euchloë Pima* and several other good species of butterflies. Mr Piazza at once discovered that Yuma would be a good place for moths (which was his speciality) and announced that he might follow me there, asking me to write and let him know how I got on, so I promised a card, inwardly resolving that I would *not* find Yuma a good place for moths. There was no need for any duplicity, for I found practically nothing at Yuma, and in the Arizona Hotel, of a strictly commercial character, I passed many a lonely hour.

However, I found a good horse in a corral in Thirs Street; black, with

46

white fetlocks, slender and well made, and with quite a good canter. How different everything is from the standpoint of a middle-aged woman. In the old days, when I might be starting on horse-back from some hotel, half the establishment would turn out to see me mount, while waiters would be running about with chairs; now I simply went to the corral, fetched the horse myself and hitched him up outside the Arizona Hotel, while I brought down my saddle and saddled him up myself, not a man standing by offering to lend me the slightest assistance or apparently taking the slightest notice of my proceedings; and when he was saddled I would promptly mount and ride away, nobody troubling themselves about me. I must own I found this way much more to my liking, for if there is a thing I hate, it is being fussed over. I enjoyed long rides into the desert, though the desert was parched up by the drought and I got no butterflies.

I moved on to Phoenix, where I had been given an introduction to a Dr Parker who took me two or three times for expeditions out into the desert in his little Overland automobile. It made me discontented with horseback ridings; the distances we covered, the rough tracks we traversed, where the little Overland would jump like a live thing over all obstacles, only to go humming on its way undaunted, and the excitement of rushing madly across that wonderful desert, was a new experience for me.

The local newspapers, for want of anything better to write about, had spoken of me as if I were quite a celebrity, but I felt it was the limit, one night past 11 o'clock, when I was in bed and more than half asleep and a ring came at the 'phone in my room. I got up feeling very sleepy and shouted 'halloo' down the receiver only to catch my own name in a man's voice which I didn't recognise, and the next word I caught was 'Republican'. Surely, I thought to myself, they aren't ringing me up at this time of night to ask me if I am a Republican or a Democrat? I had not the slightest idea which I was. However, when I began to collect my scattered senses, I gathered that the person speaking to me was a reporter for the *Republican* newspaper. I began to explain that I had been out riding all day, intending to be excused any further communications till the morning, but I was promptly interrupted: 'Yes, I know you've been out all day. I've been round to the hotel three or four times to try and see you, and I want you to tell me something about your butterflies right now, as it has to be in the *Republican* tomorrow morning.' So there was nothing for it but to gather up my sleepy senses and shout down that 'phone all the information he wanted.

I awoke on this morning, April 15, 1918, in sunny California, and I look out of my window to see the sun shining on the palm trees in Whitley Avenue. The greatest battle in the history of the world is now raging, and one feels

47

so acutely how little one can do, only stand and wait; I just work by day for the Red Cross, making bandages to mend broken soldiers, and it seems so small a thing to do.

Early in the morning I go down to Holly Springs Canyon to get food-plants and make one or two fugitive attempts to catch *Papilio Rutulus* as he floats majestically by, but go home with an empty pocket-box. When I get back I feed the caterpillars, the 100 *Melitaea Chalcedon* and *Vanessa Antiopa* larvae, besides several nettle feeders, which I think are *Pyrameis Atalanta*. Then I get a streetcar on to 9th Street, have a frugal lunch at the Waffle Shop near the corner of 9th and Broadway, and go to the Red Cross, where I baste a pile of abdominal bandages, a difficult thing I feel quite proud I am able to do.

How much more I should find it to my liking to be a despatch rider, how much more suited I should be to help – which I believe many women are doing – in military stables. But I am not in England, so I 'do the work that's nearest', though it is intensely distasteful to me. Later on a photographer came to take a picture of us all at work, which is to be published in the *Los Angeles Times*. At first I thought I would keep out of it but on second thoughts I decided to be in it, so I took one of the abdominal bandages on which I was busy at the time, and went across to sit down amongst the others, though I took care to be well in the background. Of course, as 'just a worker' I am under orders, and I often think of those unfortunate women and girls who have to spend the greater part of their lives working in factories, day after day, with no prospect of change, blindly obeying orders.

FIVE

—◈◈◈—

Butterfly Harvest

*Earning a living collecting spiders and butterflies —— a
job in Pasadena at 25 cents an hour —— Americans and
snobbery —— catching the streetcar to work —— frost
and smudgepots in the orange groves —— romantic
connection of two doctors —— the hermit in the village
of Palm Springs —— forest fires and Reds under the
beds —— a new phenomenon, car-driving wives picking
up husbands —— sailing for New Zealand*

Although Miss Fountaine does not descend to detail, her secure
Edwardian investments were not keeping pace with rising taxes
and prices. By the summer of 1918 she was almost penniless, she
says. She rose cheerfully to the challenge. While life's daily irri-
tations frequently brought out her gloomy melodramatic worst
('only a few more steps to stumble onwards in the dark, and then
– lie down and die!' she wrote one day when she had twenty years
of life and much eager globe-trotting ahead), this lack of money
invigorated her.

It was good to work and find that I *could* make a living for myself out of
my much despised entomology, though it was by no means only butterflies
that I had to collect. I had an order from Ward's Natural Science Establishment, back East, to collect for them four dozen trapdoor spiders with their
nests. I did not know a thing about spiders; where to find them or how to
look for them. However, having heard from the ladies at the Red Cross that
these creatures were to be found in the grounds of the Raymond Hotel at
Pasadena, I wrote to the proprietor and suggested I should come and get
some spiders from his grounds, and that we should share profits. He promptly
replied that he accepted my proposition, and fixed on the day following for
me to come to his place. So I went; and to my astonishment found that the
Raymond Hotel was little short of a palace.

On alighting from the street car, I made my way, dressed for the occasion,

49

i.e. spider hunting, up the long drive which wound through spacious pleasure grounds to the hotel. Mr Walter Raymond (proprietor of this costly edifice, a very important looking person) advanced to meet me and brought forward a little wizened-up old gentleman whom he introduced to me as Mr Tuttle, adding that there was practically nothing about spiders that Mr Tuttle did not know, to which statement the little, wizened-up old gentleman modestly assented. We then all three proceeded to the golf links (the Raymond Hotel has golf links of its own) and Mr Tuttle got to work, with the result that we dug out just about one dozen spiders with their nests, which he showed me how to find, and also gave me many useful hints on the subject such as how much of the nest would be required by scientists, and also how to truss and stuff the spiders. In fact I profited hugely by Mr Tuttle's information and returned that evening to Hollywood greatly satisfied.

After this I had no difficulty in completing the order from the tops of the Hollywood hills. Finally Ward wrote and told 'Mr Fountaine' that they much appreciated the way he had sent them these spiders with a promise of further orders for the future. Having promised to share profits, I wrote to inform Mr Raymond that I had received $4.20 for the one dozen spiders I procured from his grounds, and I therefore enclosed a cheque for $2.10. It seemed incredible, but this man, owner of all the magnificence I have already described, took the money – and I don't believe Mr Tuttle ever saw a penny of it.

Now Mr Newcomb of Pasadena wanted me to collect *Exilis*, and I had undertaken to get him 5,000. It was no easy matter, but, of course, the long years I had spent in the study of entomology were a great assistance, and I would gather up these tiny butterflies by the hundred, where most people would scarcely have seen one; a purely mechanical process it became at last, the only joy connected with it being in the numbers of small children that would collect around me. The bigger boys would try and learn what I was making at this business: 'Not much' was about the only answer they ever got, but I loved the small children, and I would hear them say: 'Here comes that butterfly lady again!' It was wretched drudgery at best, scorching days when I would be six or seven hours out collecting *Exilis* and never really know how hot it had been till I got back in the evening to the Wilcox Inn to find some of the guests in a state of collapse from the heat. I got my 5,000 *Exilis*, and refused to undertake the other 5,000, though much urged by Mr Newcomb to do so.

In the middle of October I left Hollywood and went to stay in Pasadena to take on regular work for Mr and Mrs Newcomb. I quite enjoyed the work, which was a good deal of handling butterflies, to which I was accustomed, and a good deal of washing glasses and fixing up the cases to contain them, to which I was not accustomed, and the rest was more or less artistic work in which I succeeded best of all; and did the very best I could for my two bits – 25 cents – an hour; and I think I gave satisfaction, while I was making sufficient income to keep myself and allow my banking account to stand practically untouched. And Charles, my dear Charles, when I told him of my financial difficulties had sent me a draft for $25, saved out of what he was making with his little shop.

Mrs Newcomb was a charming as well as an artistic woman. Sometimes rich Americans from 'back East' would arrive in their automobiles and she in her most charming manner would show them all her plaques and trays, and they would often purchase largely and leave orders that we scarcely knew how to execute with sufficient promptitude. I always did the milkweed backgrounds now, as I had attained considerable efficiency in this branch of the work and Mrs Newcomb was glad to get all she could done by me, before I left. I used to be amused sometimes at the attitude of these rich customers, ladies clad in costly furs and with manners calculated (by themselves) to befit the exalted position wealth had raised them to. Some would evidently consider that it was beneath their dignity to address more than a few passing remarks to the 'woman assistant'. Others of the better bred kind would be most kind and condescending; while it interested me to act my part, speaking only when I was spoken to and taking care to assume a manner as became one in my dependent position. How great a change would have come over some of these wealthy ladies had they known that I could claim relationship to some half dozen families in the British aristocracy. For the Americans, in spite of their democratic ideas, are the greatest snobs on earth. When these important persons had gone, Mrs Newcomb and I would have a good laugh over it, as we worked away to execute the orders they had left.

When the Armistice was signed I sat on the steps of Hal Newcomb's shop in Pasadena and one of my first thoughts was, 'and now Khalil will be able to get his passport and come to this country'. But Charles' next letter told me that a new trouble had come to him – my poor Khalil who had never done any harm in his life, the cleanest living man I had ever known; how cruel that such an agony as this should have come to him, through no fault of his own.

What was Charles's 'trouble'? Miss Fountaine leaves us to speculate. She is more specific about the disease which now began sweeping the United States . . .

'Spanish Influenza' was stalking across the lands. Schools were closed,

51

churches, theatres and all places of amusement, and people were dying all around us. Virginia, the only child of Mr and Mrs Newcomb, had it, but luckily it went no farther amongst us; and the child completely recovered as children generally did, it being apparently men and women in the very prime of life who fell victims. My friend Miss Riddell was down with it, out at Rodondo where she was being neglected cruelly, so I furnished myself with a gauze mask and an extra big supply of eucalyptus oil and went over there right into the thick of it and brought her away in an ambulance to the hospital in Los Angeles, and I believe I saved her life.

It was a terrible winter in California, the ground covered with a white frost, the oranges frozen in spite of the smudgepots which were lighted night after night for their protection; but on the whole, I look back with pleasure on those thirteen weeks at Pasadena, when for the first and probably for the last time in my life I ranked amongst the humble workers of the world. The worst part was the long cold wait in Pasadena, after my meagre dinner at a restaurant, for a streetcar back to Hal Newcomb's shop, and then going back in the dark to sit all the evening in Mrs Jagger's kitchen, getting what warmth I could out of her gas-stove. I liked this woman, but I would have liked her a deal more had she been less stingy about lighting the fire in the sitting-room.

When I returned from Pasadena I joined a party of Los Angeles entomologists to go to Palm Springs in a 'tin Lizzie' which Dr Comstock had borrowed from a friend. The party consisted of a lady physician, Dr Lord; Hal Newcomb; Mr Simms and myself, with Dr Comstock the only one who knew how to run a car, acting chauffeur. The trouble with that machine was incredible. We never reached Palm Springs that night, and while at the hotel in Banning, where we were all having dinner, Dr Comstock announced he had taken a violent chill. Dr Lord at once took him under her care, nothing loth, for there was no doubt about the relationship between these two young people; unfortunately there were two 'just causes and impediments' in the way – one being Mrs Comstock and the other Mr Lord. However, in California these little difficulties are soon overcome, and no doubt will be in this case also.

Dr Lord gave us the somewhat disquieting information that his temperature was just over 103°, and she was pretty sure that it was the influenza. She got him to bed and phoned to Los Angeles (about 100 miles) for an ambulance to come next day to take him back. We prepared to get through the night at Banning as best we could, taking it in turns to attend Dr Comstock, now in high fever; Dr Lord said he could not be left alone.

Next morning Dr Lord was left behind with Dr Comstock – I offered to stay with her, but this she most emphatically declined – and it was decided that Mr Newcomb and Mr Simms should have a driving lesson off a man from the garage here in Banning. Mr Simms acted chauffeur now, and with wonderful ability, considering that he had never run a car in his life before, and the road was none too good. But we 'made it', and got all right to the village of Palm Springs, where after some difficulty we got accommodation for two nights.

There was no mistake about the heat here, a real desert atmosphere, with a cloudless sky above, and plenty of scorching sands below. We were, however, still eight miles from the Springs, so that there was nothing for it but to shake up our old car and prepare for the worst, for the road was pretty bad all the way. However, bar being stuck in the sand once and being very nearly overturned several times, all went well, and we arrived.

There was scarcely a butterfly to be seen, and I devoted my time to the numerous dragonflies. But it was a wonderful place we had come to, the water clear as crystal, and the rocks bare and almost devoid of vegetation except for the palms. We got up what interest we could in the Hermit, a rather artistic-looking young man, not more than thirty, with large innocent blue eyes and a beard, and long fair hair which hung in soft ringlets over his shoulders; some of the women seemed greatly interested in this creature, but the men got him on their nerves. . .

Early in 1919 Miss Fountaine was in Hollywood, now 'with its atmosphere of Movies, Movies and still ever Movies'. She cabled Khalil $250 for his passage money and while she waited for him, visited the Yosemite national park, then just beginning to feel the burden of what Miss Fountaine still sometimes wrote as the Auto-mobile. She, the traveller who had been unimpressed by Niagara, thought the park 'stupendous – I had never seen anything like it'.

But man has discovered this wilderness of beauty, and amidst the grandeur and splendour canvas tents were everywhere, crouching in groups beneath the shadows of the gigantic pines, evidences of man's insignificance one could not get away from. The nights were bitterly cold, and in Camp Curry a huge bonfire would be lighted; and every evening after dusk shouted messages would be exchanged with an hotel built on the edge of the cliff, 3,000 feet above us, so clear was the atmosphere. Then a great red, starry light would fall down the face of this cliff while all this immense crowd of human beings from above and below would join in singing a verse of one of America's anthems; and still the old trees looked down, and watched it all.

She hunted Yosemite butterflies, before joining friends who

'to crown it all, invited me to return with them in their Cadillac Auto-mobile; which was indeed a delightful treat'. She returned to find letters from Khalil, who announced the return of a long-buried skeleton; he had received a letter from his mother, and in this letter he had been told that his wife was still alive, 'a thing I was not at all surprised to hear, as I never had thought she was dead,' remarked Miss Fountaine. 'But surely at the end of some twenty years' separation, by no law of God or man had this disreputable woman any further claim upon him?'

She pursued *Vanessa Californica* on Mount Shasta, near Sisson, and noted that 'the forest fires were terrible this year, immense areas blazing all the time, and no sooner had the fire-fighters got one under control than a fresh one would break out. It was no doubt the work of incendiaries, probably the I.W.W.,* but it never could be brought home to them, so the wholesale destruction went on and valuable timber perished with the rest. It seemed the whole country would soon be a blackened wilder-ness. It was a cruel sight to see those areas which had escaped the fires of other years now being burnt and laid waste on the immense scale that everything is done in this great, colossal America. For hours, sometimes all day, the sun is darkened with the smoke, and here in sunny California its red orb is seen as through a London fog.'

Miss Fountaine also observed a new sociological phenome-non: 'The Sisson ladies would go round in their automobiles and pick up their respective husbands from time to time; I used to tell one of them that she and friends were always in a chronic condition collecting husbands. As I sometimes accompanied them on these "collecting expeditions" I saw the country well.'

More letters from Khalil postponed their meeting again: 'He was *not* yet cured of that horrible disease; as is the case with many men who think they are cured, it had shown signs that it was still there, and this had been confirmed by the opinion of two doctors.' She decided to sail south for New Zealand as quickly as possible, both to be nearer Khalil and to avoid winter, with increased risk of influenza, in the northern hemisphere. 'All sorts of regulations and forms, unheard of before the beastly war, now had to be gone through with. Of course, my passport had to be fixed up, but what was even more difficult was to get a berth on a boat. Every boat was booked for months on ahead, and my only chance was through someone cancelling. However,

* The I.W.W. – Industrial Workers of the World, the 'Wobblies', a socialist trade union group – were a favourite bugaboo of respectable America at the time.

I handed them my cheque for a round-trip first class from San Francisco to Wellington, and back from Auckland to Vancouver, available for one year from date of sailing, and in December 1919 the *Sofia* sailed at last, past the Golden Gate and out to sea.'

SIX
❖❖❖

Heading South
1919

*Purple prose in the tropics —— in Fiji at 58 and at peace
—— the English boarding-house abroad; 'Kenilworth'
and 'Victoria View' in the South Seas —— Khalil but
still no wedding —— back to the US and good hotels at
two dollars fifty —— England on the* Aquitania ——
philosophical reflections on life's third-class passengers

The diary for the year ending on 15 April 1921 began with a
stretch of descriptive prose which edged, as Miss Fountaine was
inclined to edge, into the purple: 'Beneath the warm glow of a
tropical sky the languid air hung lazily above the soft gentle
murmur of a calm and lovely sea, where the grey misty outline
of jungle-clad mountains threw their soft shadows upon its
glittering surface . . .'

It was 16 May 1920, a Sunday morning, and Miss Fountaine's
58th birthday, and she reflected, as she sat on the balcony of the
Grand Pacific Hotel in Suva in the Fiji Islands, that she was
almost at peace, because 'I knew that youth with its mighty
passions was dead at last'. In this she was mistaken.

She rapidly moved out of the Grand Pacific into Waverley
House, a boarding house, where she says she was more happy
and comfortable. The longer and more distant travels of her
later years were putting a strain on her purse which the investment of Sir John Lawes forty and more years before could not
meet without that kind of economy; and though the deaths of
some of the original seven beneficiaries from Uncle Edward's
will had increased the share of the survivors, there was not
enough to cover globe-trotting, post-war Income Tax, and
hotels as well. Miss Fountaine does not record what her income
now was, but it put her into the Waverley House class, not into
that of the Grand Pacific.

These boarding-houses, which offered economy when she and Khalil were travelling together, and economy and company when she travelled alone, are to the modern reader one of the surprises of her travels. That she should find an establishment called 'Kenilworth' in New Zealand is perhaps to be expected, but that other boarding-houses, their names straight from the greyer English seaside resorts or Sir Walter Scott, should be found in such far-off and romantic places as the Fiji Isles, Burma, the West Indies and Hong Kong, is a sudden reminder of the culture which followed trade and the flag.

Although she was not well – among other things she had an ulcerated leg, possibly one of the infections which were common in the South Seas – she took a sugar company train half-way across the island to spend seven hours a day wandering in some of the remaining natural bush with her net. When the Fijians became too childishly curious about her collecting, 'I found a capital way to get rid of them; just to have a fit of coughing (I had a bad cold) would send them off in a hurry – the influenza had been very virulent last year in these islands, and many Fijians had succumbed to it. All the same I was grateful to an old Fijian and his son who brought me a green cocoanut and cut it open for me just when I was wondering how I could endure my thirst till I got home.'

She remarked that the Fijians were only fifty years from cannibalism and the vilest cruelty and enslavement by their chiefs; now 'the Englishman who wrought their deliverance from slavery is in many cases nothing but a slave himself, to the demon Drink. I have rarely seen such drunkenness, even in India and other British colonies . . .' She visited the leper island of Molakai, spending three weeks with the island's doctor and his family before sailing on to Auckland, New Zealand, where she arrived on 10 November.

The day after my arrival, when I was down in Queen Street, that busy noisy thoroughfare, suddenly an explosion rent the air, immediately followed by absolute silence; not a sound and not a movement, as though the whole city had suddenly become paralysed. The trains had ceased to run, every cart or motor-car in the street stopped, passers-by were motionless. I was in a shop at the time, but the man serving me was standing just exactly where he had been when that blast had first rent the air. For a minute or two I also remained silent, then made a slight movement, as though I would address the man in the shop; but he waved me to remain silent, and then I knew: it was 11 am, and today was the 11th of November, and this great silence was in honour of the dead.

It was two years after the 1918 armistice. Miss Fountaine thought the silence lasted for five minutes; it would, in what became tradition, have been only two. 'Then the city sprang into life again, and the graves in France were once more forgotten.'

She had arranged to meet Khalil in Wellington; 'scarcely a place to be selected for its climate, about the most windy place on earth,' she remarks, quoting the old New Zealand tale that one can always tell a Wellington man away from home because at every street corner he will instinctively put up his hand to hold his hat on. She filled in the time before she was to meet Khalil by visiting the lakes and boiling springs at Rotorua, and by collecting *Pyramis Gonerilla* from giant bush-nettles. Then his boat arrived. Their meeting was something of an anticlimax.

I saw others meeting their friends and I watched eagerly to see Khalil's dear face once more, but it was not till I had left the gangway and walked a short distance down the wharf that I did catch sight of him at last; he was standing alone, looking over the rail of the boat, not apparently in the least on the look-out for me; he did not look at all ill, only very very sad, and it was a few seconds before I could attract his attention. Then he smiled and pointed to the gangway, meaning, I suppose, that we should meet there; and I felt a curious feeling not exactly of disappointment, but of something quite different to what I had expected. It was not possible after those long years of separation that we could either of us ever be quite the same again. However, we both agreed now that the sooner we could be married the better; the other boarders at Kenilworth took quite a lively interest in the matter, and I received congratulations. I intended the wedding to be absolutely private; indeed we had practically decided to be married at a Register Office, and then just slip quietly away that evening for the South Island.

Charles was wonderfully better than I had expected to find him; he could take quite long walks, though we spent a good deal of time sitting in the Botanical Gardens. It was there, only two or three days after we were together again, that he began to tell me how he must go and see his mother; it seemed she had been writing to him, reminding him that she was a very old woman now, and if he put off going to see her any longer she might be dead. I felt hurt; we had hardly met before he was planning to leave me again, and I raised a strong protest against it.

However, he lost no time in going down to the town to make arrangements for our marriage; and that it was to be a marriage only for mutual companionship and of mental affinity was all I wished for.

Then he came back one evening: 'No good! No good!' he said. 'They asked me at the Register Office if I had ever been married before, and I had to say I had, and that my wife was still alive. And though we have been separated for over twenty years they said it was no use, until I got a divorce.' (From what I could gather, he could easily do that.) Then he went on to say, 'When I said I had been married in the Greek Church they told me that I cannot get a divorce here, I shall have to go back to the country where I was married. So I will go back and see my mother, and fix up this business at the same time.'

I thought, 'Go back and see his mother' – that is all he seems to care for! I felt he no longer cared for me as he had once done. But he seemed very much disappointed; so was I, and of course explanations had to be given in the house; we just said it was because 'Mr Neimy belonged to the Greek Church' and there was no Greek Church in New Zealand. Which was true enough, as far as it went.

We both tried to make the best of it, and to enjoy at least the time we were to be together. It was good to have Khalil with me, and when on Christmas Day we went to the service in St Peter's Church, and knelt together before the altar, I did feel a great thankfulness to God, who had brought us together once more. Often Charles would tell me about his life in Sydney, and the strange experiences he had in that little shop in Kent Street. I would shudder to think of the dangers to which he had been exposed, for Kent Street is practically in the slums, so that he had been up against all sorts of people, fighting his way through, single-handed, ill too, and lonely.

I was ready to take every possible care of him, and to give him anything he wanted, as far as it lay in my power, to try to make it up to him for the miserable time he had spent in Sydney. So we went away together as already planned to the South Island, only it was not a honeymoon trip, no rice and old slippers after all, but the old, unsatisfactory relationship, travelling together for the sake of collecting entomological specimens, had to be resumed; all the old subterfuges, falling back upon the butterflies as the reason for our travelling together in this way; and again, as in the old days, the butterflies saved the situation.

We collected *Antipodum* on the tussock-clad hillsides near Springfield, and a tiny *Chrysophanus, C. Boldenarum*. The deep violet of the males made it most attractive; besides it was such a tiny little thing, and Charles always loves little things; indeed his disposition had become so gentle and kind as the result of all he had suffered, that I could see he felt very tender-hearted about killing butterflies at all now, and only did it to please me. But neither of us were the same; that fearless independence, that scorn of danger, was no longer ours; for instance, both of us now were afraid of cows. In South Africa

Khalil would face bulls with an intrepidity that even I held to be somewhat risky; now when a bullock began racing up the hillside towards us, why, we both by common consent turned and fled.

While they were hunting butterflies near Nelson, she tried a little meddling with nature in the high-handed Victorian way, putting out two dozen larvae of a Californian butterfly, *Papilio Tolicaon*, 'at the roots of the fennel which grows in such rank profusion in both islands. I wonder if they will ever hatch, so that Californian butterfly may establish itself, and if so, will Dr Tillyard's fears that they may increase to such an extent as to make serious depredations on the carrot and parsnip beds be not without cause?'

Even had she taken these fears lightly, Miss Fountaine was too experienced a naturalist to be excused for having risked it; she had just remarked on how the introduced blackberry had increased amazingly in New Zealand, and a little while before she had seen that the introduction of outside birds and a species of Indian wasp had wiped out butterfly populations in some Pacific islands.

They returned to Wellington to wait for their boats, his to Port Said, hers to San Francisco. Miss Fountaine must have been trying to persuade Khalil to ignore the calls of filial duty and probably overdoing it; she records that 'one day he said in a tone of considerable irritation, "You really must leave me alone, if I wish to go home to my mother. Then I will fix this business and we will meet again in London." So I gave in.' He had to stop in Sydney to pick up some money from the sale of their property in Myola, 'Charles having very wisely sold the place about a year ago, and only just in time, for the very week he received the cheque a cyclone had blown the house away.' It was the end of their Australian disaster.

The New Zealand visit had not been a comfortable time for either of them, for Khalil seemed to resent her efforts for him, and she was wounded by his attitude. They were distant when the parting came, 'till the call "All visitors on shore". Then we were lovers again, kissing each other goodbye several times over, and as the *Marama* began slowly leaving the wharf I could see tears in his eyes . . .'

It was winter when Miss Fountaine landed in San Francisco, and she crossed the United States as a tourist rather than as a butterfly hunter. She found the warmth of American trains comforting, the pleasure of a room with private bath ('that most desirable luxury only, as far as I know, to be found in American hotels') not excessively expensive at two dollars fifty a night; a light dinner ordered without benefit of her spectacles to check the menu turning out to be shockingly costly at a dollar seventy-five. She remarked, with curiosity rather than with condemnation, on girls in the new fashion of 'short skimp skirts almost up to their knees, with no protection but silk stockings from there to their shoes; then they were clad in heavy fur coats but had low cut necks. I cannot imagine what it can feel like to have part of one's person so bountifully covered and the rest left almost wholly unprotected against this bitter cold'.

She visited Salt Lake City, and wanted to know whether any of the several widows of Brigham Young, the Mormon leader, had married again. (Her guide seemed embarrassed by the question, and said 'No' somewhat shortly, she noted.) She celebrated 15 April 1921 in Salt Lake City, and then moved on to Chicago.

Of all the plague spots on the face of this earth the city of Chicago is about the most virulent. I disliked the place when I visited it before, but I loathed it now; if all Americans were like the people of Chicago then would they be the vilest nation on earth, instead of one of the very best. I was nearly knocked down there one day by a young man in a crowded street, and he never even turned to see what had happened, but just went on still forging his way through the throng – and, by the cut of his coat, he might have been a millionaire. Then too, the Irish movement, so injudiciously taken up by Americans, was much in evidence. A huge meeting on behalf of that nation held in Morrison's Hotel, where I was staying, did not tend to lessen my mental discomfort. It got on my nerves to see well-clothed American women the next day eyeing me with looks of righteous indignation on their well fed faces; I knew they had swallowed enough lies at that meeting the night before to upset their moral digestions for some time to come.

In the midst of all this, there was Mr and Mrs Goodspeed, two people I had known in California, and they were still of the West, and as kind as they could be; how they could bear to be living here after Hollywood, I can't think.

Before I left San Francisco I had booked my passage by the *Aquitania*, sailing from New York, on May 3, and after a decidedly pleasant voyage on this monster liner we reached Southampton on the 10th. It was nearly nine years since I had seen my native land, and I turned away from the first sight of the coast of Cornwall, feeling I should have cared nothing had I never seen the old country again. I only longed for rest, just to get home.

61

After seven relaxing days in the comfort of the *Aquitania* she still felt tired, but had no sympathy for the porters on the quay at Southampton who 'crawled about, shifting the goods and luggage in a kind of dilatory lazy manner which showed me the sort of thing to expect back here in my dear native land. "That's the pace that's killing England", remarked one of the men passengers who with the rest of us was leaning over the rails . . .'

One cannot disguise the fact that Miss Fountaine, with the years, was becoming a Blimp. Twenty years before she would have seen the humour in those last few words. When she discovered that the woman in whose charge she had left her studio had not failed her trust in nine years, though this custodian was now 76, Miss Fountaine's response was, 'Such loyalty is rare in these days amongst the working classes.'

And – though this is to anticipate a little – when she economized by sailing third class on a subsequent journey, she selects the word 'humiliating' to describe the consequent inconvenience. Disembarking in France, 'I went to the Hotel des Ambassadeurs and tried to forget that I had just come off a boat as a third class passenger; it was good to be shown to a place in the Salle a Manger by the Maitre d'Hotel with every mark of civility and respect. I thought with feelings of the profoundest pity of those who are always third class passengers all through the journey of life, but then they have not sprung from a long line of ancestors who had ever been born to rule and to receive homage from their fellow creatures, so maybe I might just as well have reserved that pity for myself.'

This is all good knockabout stuff, and one might suspect that Miss Fountaine had her tongue just a little in her cheek. But atrophied imagination can be the only excuse when she says of a friend, 'I was surprised that Mrs Spens was greatly in favour of helping the Russians to save their children from starvation in consequence of the terrible famine . . . these children will merely grow up to become a scourge to the whole world, cubs of Bolshevists.'

On the other hand she was shocked by the sight of unemployed ex-servicemen busking – or begging – in the London streets; shocked at the price of war, and pertinently asking, 'what would the men say if it were possible for a lot of old women to bring about conditions which would mean the wholesale slaughter and mutilation of all the youngest and most perfect of the girls and women? They would never tolerate it. Then why should we?'

SEVEN

Reunion in the East
1922

*Khalil ill —— reunion and time 'on swallows' wings'
—— Uncle Lee Warner, prayer book in one hand,
picking pockets with the other —— off to Rangoon
intoxicated with happiness —— whales and flying fish,
dak bungalows and golden pagodas —— English soldiers
with well-washed red necks, and Indians spreading
sedition —— staying with circus folk —— Siam and
stuffy British diplomats, friendly Americans —— 'No
young girl could have a more devoted lover'*

In London she had found other changes; her former landlady
ruined by the war, and the boarding-house no more; she met
Geraldine, declaring to her that she intended to marry Khalil
and that Geraldine could either accept a brother-in-law or re-
nounce a sister; they were friends again 'after a little skirmish-
ing'. She visited an entomologist friend and was lured into playing tennis, acquitting herself well – there were young people in
the house and such company always stimulated her. She served
on a jury at the Old Bailey, 'a nuisance from which women had
hitherto been exempt but now had to submit to in consequence
of being enfranchised'.

But Khalil had been ill again, with a recurrence of malaria,
and unable (or unwilling) to respond to her increasingly anxious
cables urging him to join her in London. Heartbroken by his
letters – he had fever every night, he had lost almost 70 lb in
weight, goodness knew when he would be well again so they had
better break off their engagement – Miss Fountaine declared she
wouldn't dream of abandoning him in his hour of need, and set
out for Beirut, a ministering angel perhaps not wholly welcomed.
She could not help remembering the last time she had sailed
there, when the interfering missionary Mr Segall saved her from
a bigamous marriage to Khalil, or, in her words, had made her
visit 'such a failure' – an odd description. However:

. . . there was no one to interfere with me now. I found myself again walking through the filthy streets of Beyrouth with all the fearsome sounds, sights *and* smells of the Near East, together with the flies and the jargon of many tongues.

It was New Year's Day, 1923, when Charles was to arrive; I had lived through those moments in imagination, expecting that I would see him helped out of the train, a mere wreck of his former self. I was almost turning away, believing that he had not come, when I heard his voice behind me. And I turned to behold Charles, in a Syrian tarboush, awfully thin and drawn but not nearly so pale and weak as I had expected. The very sight of him did me good, but: 'Mother she is not at all well, and after one week I have to go back to her,' he said. I had gone through all this for practically nothing.

Of that one week it is enough to say how happy we both were to be together again, how we poked about in the old streets, and spent a few piastres in the bazaars, so that the time passed away on swallows' wings. Charles had not succeeded in getting a regular divorce from his wife, a legal separation was apparently all he could get. This did not trouble me much; we could marry just the same if we wanted to, as things are in the world now, but did we want to? I very much doubted if Charles did, and I felt almost convinced that I did *not*.

I had not enough money to return to Europe, so I took a boat to Haifa in Palestine. Haifa is at the foot of Mount Carmel, and during the four weeks I was there I went up that mountain many times, sometimes alone, sometimes in the company of some other English woman, and Charles wrote me some delightful letters and said many dear things to me, so I was quite satisfied.

The difference in Palestine since the British occupation is very remarkable; there were railways and roads, motorcars, British soldiers, and sanitation in the hotels and many other things conducive to comfort. It was much too early for butterflies, so I got busy turning up stones on Mt Carmel looking for scorpions to send home to Mr Main (a collector); loathsome things I had hitherto always avoided. But Mr Main is most kind in taking care of my collection at home. There are plenty of stones on Mt Carmel; too many in fact, as it made the chances of a scorpion being under those I turned over somewhat remote; albeit when I looked for a much-needed rest I would fear to sit down, in case I should happen upon some stone which I might for the purposes of collecting have overlooked!

There were some interesting places in this neighbourhood, one of them Acre, which brought to my mind the Siege of Acre, somewhere in the forties of last century, where my Uncle Robert Lee Warner, as a boy in the Navy, was said to have distinguished himself. He subsequently, I believe, returned home, and gave it up as a profession, having suffered so horribly from seasickness that he declared he would 'rather break stones on the roads'. But he had left his small footprint on the sands of time at Acre; though he did spend the rest of his life with a prayer-book in one hand while he picked the pockets of his poorer relations with the other. But I digress.

Miss Fountaine was back in London for the end of her diary year. It was a cold day; she recalled in the detail with which she usually described that extra significant day how a bitter wind made her window rattle and penetrated every crack as she breakfasted off nicely crisped bacon supplied by a Mrs Bird (whose duties also appeared to include showing-in of visitors). She spent the morning drawing the *Aporia Crataegi* caterpillar – 'quite a success, but it was a pity I didn't stop there instead of going on to paint one of the pupae of that same butterfly, as I did it badly . . . I looked up the specific name of the arbutus and began printing the little note under the larva of *C. Iasius* I had drawn at Hyeres.' All these are in her sketchbooks, preserved now in the library of the Natural History Museum in London.

She had 'a simple lunch – of cake and wine and gingerbreads', smoked three cigarettes, wrote letters, arranged some of the specimens in her display cases more neatly, and dined that night with an acquaintance made on her travels, at a women's club in Sackville Street, Mayfair. She disliked the crowds and the traffic of London Streets – it made her long for the darkness of nights in the jungle, she says – and retired to her 'cupboard-sized bedroom' with some pleasure. Such days were a quiet and civilised preparation for another long journey.

Early in August she booked a passage from Liverpool to Rangoon for herself, and one on the same ship from Port Said to Rangoon for Khalil. In spite of her cool and dry-eyed parting in Beirut, their reunion at Port Said was enthusiastic. She had been in her usual state of agitation when Khalil did not appear the moment she had first looked for him; needless anxiety.

Suddenly all my fears were at an end, on the deck I saw his dear, familiar figure, his back turned, bargaining with the porters who had brought him and his luggage on to the boat. We have often parted, and met again, but I do not think any meeting had ever been quite like this one. I felt intoxicated with happiness and contentment, and now the rest of the voyage was a continual joy.

According to some of the passengers all sorts of terrors were in store for us in Burma. We were to be crushed by boa constrictors, chased by king cobras (who could travel at a greater speed than a horse can run); of course we were almost certain to be eaten by tigers, while even the tame buffaloes were a terrible menace to white people; and as to the wild boars they were so savage that to meet one in the jungle would mean certain death to us both, and moreover *all* the wild elephants were 'rogues' in Burma! Well, we knew enough of these sort of yarns, and also of the countries we hear them about, to know that while listening it were just as well to have a pinch of salt handy. So, unintimidated, we watched for whales spouting in the hot seas beyond Ceylon, and the flying fish as we passed within sight of the rain-steeped Andaman Islands, till at last on a murky, hot morning we steamed between the flat, green shores of the Irrawaddy River up to Rangoon, a huge, dirty, straggling place, hot and moist, with the usual riot of colour in all its main streets, and the usual terrible squalor in all its native quarters that one finds in every Oriental city.

There is always a pleasing novelty attendant on the first railway journey through a country hitherto unknown to the traveller; thus Charles and I were greatly interested when we left Rangoon behind, and penetrated into the country beyond, which at first was miles and miles of flat rice fields, of the most vivid green, with the so-called ferocious buffaloes much in evidence. The native villages often seemed as though they were built on purpose over fearsome swamps, and how the strongest of the inhabitants of these gruesome hovels ever survived such horrible conditions, was a marvel to which there seemed no explanation. But what interested us most were the immense numbers of golden pagodas, on the summits of hillocks. By and by when it was beginning to get dark we could see that at times we were passing through jungle, into which we peered in vain in the chance however slight of catching sight of a tiger, which of course we never did.

There was no longer any hotel at Thandaung, but we were greeted at the American Mission by two charming American women, and for five rupees a day each we were put up at the Mission. We boarded with them in the big dining-hall, with eighty pupils of the Mission School. The noise was tremendous, especially at mealtimes; it was like a huge, happy family which for the time being Charles and I had become members of.

Though there was any amount of jungle, we did not do anything like so well with the butterflies as we had hoped. The school had a glorious view right across mountains clothed in jungle to the wide plains some four thousand feet below, where broad rivers and stretches of flooded country reflected the most glorious sunsets I have ever seen. No doubt there were tigers and plenty of them, but I fancy they preferred the hot jungles of the low foot-hills and the plains to these higher, cooler altitudes, so we soon lost all fear of meeting one, and indeed *my* only fear was that we never would see one, and – we never did!

They travelled on through Burma, learning more about the varied styles of those Kiplingesque wayside halts, the dak-bungalows. At Pathechoung, for example:

Miss Amburn having procured for us a good ham and some other supplies enough to last us three days, we went down to Pathechoung. We had been told that this was a good, roomy bungalow (which it certainly was) and that it provided mattresses on the beds, crockery and cutlery, which it certainly did not. The beds were quite bare save for a stretch of coarse canvas, there was almost no crockery or knives and forks, and the one and only spoon belonged to Charles. Moreover, when night came on, and we asked for a lamp, the chokidar replied by vigorous shakes of the head, and then made as though he would go to sleep, which intimated that after dark here, all we could do was to go to bed. After considerable pressure a small, gruesome lamp was placed at our disposal, which emitted a dull, rather depressing light, just enough for us to grope about by. There were practically no cooking utensils either, which made it all the more marvellous that Charles managed to make the most delicious coffee for our breakfast every morning. But there was no doubt about the abundance of butterflies.

We spent a month in Maymyo, one of the principal hill-stations. The English people who have made their homes there consider themselves to be most highly favoured, because it would turn so cold that frosts might be looked for at night towards the end of December. This did not exactly appeal to me, but as it was now only the very beginning of November Charles and I put up at Dales' boarding establishment. We hired two Burma ponies, which, considering that they spent most of their time in gharries, were quite good, though very small. On Sundays we went to the barracks' church, which would be full of nice, clean British soldiers, with their well-washed red necks, looking just the same as when they left their villages far away in England. The collecting was quite good, especially along grassy rides cut through the light jungle of the hills, ideal places for up-country collecting.

We spent just one night in Mandalay and visited the old palace of the Burman kings and queens, the last of whom had been such a monster of cruelty that the British had deemed it their duty to depose him, and annex his country to the British Empire, for which act the Burmans of the present day are highly indignant, and nothing but their excessive laziness prevents an open revolt. What with the Indians, who are busily sowing sedition amongst them, and their own ever-increasing discontent, it is only a question of time before we shall have some very serious trouble in this country.

Miss Fountaine and Khalil added to their experience of boarding-houses on this tour: there was one run by a pair of quarrelling Eurasian women, one of whom boasted of having been married to, and been deserted by, an illegitimate son of King Edward VII. (She had found consolation, Miss Fountaine noted, in another Englishman of less exalted origin, who was relieving her of her money). At another boarding-house they shared rooms with a couple who were touring with a circus, 'their share in the public entertainment being the exhibition of a Kaffir man who, having two horns like a wild deer growing out of his forehead, brought in immense crowds and a fine income to them'. It was a long way from South Acre rectory.

By stages dictated by the length of time each of her letters-of-credit could be made to last, they journeyed on to Penang, and into Siam. The years had not lessened her ability to be interested in everything:

... how the geological formation of the numerous jungle-clad hills and hillocks, which seemed to be the leading feature of the country, bore such a striking resemblance to the island-strewn coast a few miles westward, and as Charles remarked, bore every evidence of the sea having once been here, and receded ... Near one of these hills was a Buddhist temple in a huge, natural cave right through the cliff with an opening at either end. Inside this weird tunnel was a colossal figure of the Buddha, lying with half-closed eyelids, and a whole retinue of smaller figures guarding it. From time to time some Siamese peasant would visit this sacred shrine, and we would feel like interlopers, and not a little solicitous for our own safety; but the Siamese on the whole seemed to be a friendly, inoffensive race. There were more butterflies here than in any other place; we discovered some beautiful *Euploeas*, and a few other things ...

The old awkwardness of their relationship came up again; they made the acquaintance of some friendly English people ('there are often charming Englishwomen buried alive, as it were, in these out-of-the-way places'), but Miss Fountaine was anxious to move on – 'make our getaway' – before their friends discovered she and Khalil were sharing the same room at the official Rest House. They travelled on by train, reaching the far side of the river at Bangkok in the evening.

The river was about two miles wide with a very strong current and I absolutely refused to entrust Charles and myself to a sampan. At last we hailed a launch which, though hauling a big barge and a few other insignificant details, was persuaded to draw up at the landing stage and take ourselves and our belongings to the other side. The men in charge being quite impervious to my demands to be landed as expeditiously as possible, we found ourselves drifting downstream with a persistence that I felt would end in our getting out to sea. But they were doing the best thing for us as well as themselves, for having disencumbered their tug of the barge at a wharf, they took us to the landing stage of the Eastern and Oriental Hotel miles away from where we had started. Of course, the E and O Hotel was much too expensive for us, so we left our luggage there and after some difficulty found the Europe Hotel, recommended by a young Italian who was travelling with a young Russian woman who passed as his wife.

One of the first things for us to do was to present Mr Greg, the British Minister, with a letter of introduction from a Miss Gaddum, a naturalist we had met in Rangoon. As is usual with letters of introduction, it had been left open, so I peeped in to see what she had written, and, finding it began Dear Robert, concluded that she knew the British Minister quite intimately. At the British Legation, however, an exceedingly pompous person presented himself as Mr Greg, saying, 'I know very little about Miss Gaddum except that she was a friend of my eldest sister when I was a child.' This was what the lady had told us, and I suppose she had not seen Dear Robert for many years, so how could she know the small brother of her friend had turned into so very important a person, especially in his own estimation?

Our reception was most chilling; we were both very amused, but Robert did us a good turn at last, telling us that the small son of Mr Brodie, the American Minister, was a keen collector of butterflies and suggesting we go on there. What a contrast in the way we were received by the Americans – we spent a delightful day with the Brodies, going with their two children in their automobile to the park in the morning and afterwards returning to the American Legation for tiffin.

With the help of the Brodies we located a certain Mr Godfrey, to whom, having found his name amongst the Fellows of the Entomological Society, I had written from Burma, without receiving any answer. But now he fixed up a time for us to go and see him and his collection. I have invariably found entomologists all over the world ready to help each other, but Mr Godfrey was the exception to this rule; we had to dig down, as though searching for hidden treasure, to get any information out of him. He seemed to delight in raising obstacles in the way of any plans we suggested with a view to working Siam, declaring that the only method was to take a big retinue of coolies, with tents etc., at an expenditure of about £100 a week! I presume this would include elephants. The fact was that hitherto Mr Godfrey had been the only lepidopterist of any note in Siam and he evidently desired to remain so.

Bangkok was a very interesting place to visit, and one of the most artistic-ally picturesque, as well as one of the most evil smelling places I have ever visited, both of which facts were more or less attributable to its having many waterways, spanned with numberless wooden bridges which, what with the native houses and boats, together with the filthy habits of the people, tended to charm the eye but was ever a great offence to the nose. The only collecting was in a garden, where I used to go sometimes of an afternoon with my net.

We sailed for Hong Kong in a Danish boat, the *Banka*, as steady an old tub as one could wish for, heavily laden with teak and rice. This trip was generally accomplished in four or five days; we arrived on the ninth day in a wild storm of wind; I was too ill to care much what happened . . .

> Hong Kong added to their collection of boarding-houses as well as to their collection of butterflies. 'Victoria View' was un-comfortable, and without proper sanitation, but their rooms were large and well furnished – 'too well furnished, for Mrs Ogilvie has a mania for going to sales and buying up secondhand furni-ture, so her two houses are bunged up with it. Charles in his room has no less than three wardrobes; I have a large chest of drawers and two big wardrobes, but a great dining-room table adds considerably to my convenience.' She needed little space, her butterflies and caterpillars a great deal.
>
> From the park in Hong Kong they collected butterfly eggs 'laid by a dark blue wild female, *Euploea Amymone*', which duly hatched on the morning of 15 April 1923. This was a day when Miss Fountaine reflected on her relationship with Khalil, who was now permanently transformed, in the diary, to Charles. 'It is a very remarkable thing the way this man still cares for me; I am now an old woman, and no young girl could have a more constant and devoted lover, for he still seems to have not only a real affection for me, as might be the case after all the years we have known and loved each other, but even more a passion ever seeking to find an outlet through kisses – that strange expression of human love.' Her feeling for Khalil, on the other hand, was maternal, Miss Fountaine declared.
>
> She began the day writing to her accountant in London, in-structing him to sell £300 of her War Loan, 'as money I must have if Charles and I are not to be held up here for the rest of our lives', and she ended it watching her fellow-boarders playing billiards before she retired ('Charles and I hooked our little fingers, a time-honoured custom between us') to watch one of her *Papilio* caterpillars change its skin.
>
> She began the following year's diary – that for the twelve

months up to 15 April 1924 – declaring for the first time her lack
of enthusiasm:

What was once a pleasure and delight to me has now become an arduous
effort, for which I scarcely know how to bring sufficient willpower into force;
my memory fails me, and my eyes are dim to the scenes I would strive to
describe. Not that I desire to draw too vivid a picture of the time we were
compelled to remain on under the roof of Mrs Ogilvie's somewhat *mis-
managed* establishment, which for lack of funds we were unable to vacate.
But in every place there are always some advantages, and when a spare bed
was taken away from my room and another huge dining-room table moved
in in its place, my joy was complete. For we were very busy here now and
had lots of cages full of larvae, for which this additional accommodation was
needed.

Any attempts to penetrate farther into China, even if we had had the
money, would have been only at the greatest risk; for China was in a turmoil
from end to end; trains had been attacked by bandits on the main line between Shanghai and Pekin, and many foreigners robbed, and taken away by
the bandits to be held for ransom. Even on the sea and up the rivers pirates
were busy attacking boats, while in the interior civil war was raging.

It was an unusual sight to see a Chinaman with a bag slung at his side
busily engaged in sweeping the grass with a net, and upon enquiry we discovered he was collecting grasshoppers for food. A fearful thought came to
me – did they also look upon the caterpillars of butterflies as nutritious
dainties? So one day when I had several larvae of *Papilio Clytia*, which we
had just found, I determined to decide this point at once. Seeing a grass-
hopper collector busy with his net, I went up and offered him a large, fat
caterpillar of this species, as though by way of suggesting a little change for
his menu; but to my intense relief it was declined at once, with every symptom
of disgust.

Mrs Ogilvie's houses were all full up now, mostly with a crowd of people
of all nations and colours who belonged to a large fair. Then a troupe of
Italian opera singers arrived, but these kept themselves very much to themselves. However, it was quite a good company, and Charles and I went to
hear *Carmen*, and would have enjoyed it save for the overpowering heat of
the Opera House.

> If the circus folk with their 'horned Kaffir' at the earlier
> boarding-house were a long way from South Acre, *Carmen* in the
> heat of the Opera House at Hong Kong was a still longer way
> from La Scala in Milan and the flower, messenger of an unholy
> passion, on the crimson cushion of the opera box in the 1890s.

71

EIGHT

The Professional Collector
1925

Hard beds and happiness in the Philippines —— hunting the Magellanus —— *the locals would kill us if they knew —— peg of palm spirit on a calm evening —— Hong Kong, the Canadian Rockies and smuggling cigars —— wealthy collector from Beckenham asylum —— lunch with the Rothschilds —— Charles takes her at her word, 'something no woman ever wishes to happen' —— listening to a Commons debate ; MacDonald mad, Churchill a bad speaker —— buying new studio, a lease 'to last until I was 102' —— West Africa and 'small chop' —— lost in the jungle*

Their next collecting-ground was to be the Philippines. The islands agreed with Miss Fountaine; after a slow start the collecting was good – and spectacular; some of the largest and most colourful of the world's butterflies are to be found in the islands, and they are well represented in the Fountaine–Neimy collection. She was in any case interested again in her surroundings; the diary comes to life. She noticed the pony carriages taking the place of cabs in the island, and sprang to the defence of the island's poorly treated ponies, which had 'added to their other miseries the cruel torture of bearing reins, for which no doubt the American occupation was responsible. Charles used to go for the coccieros – drivers – compelling them to get down and take it off, or at least loosen it. Any unnecessary use of the whip would be speedily put a stop to, possibly by a poke in the coach-man's ribs from the end of my butterfly net.'

They moved about the islands, nets in hand. 'One day Charles caught a young female of *Ornithoptera Rhandamanthus* and on being placed in a cage on my verandah with a good supply of aristolochid leaves, she began to lay that same afternoon and deposited about two dozen ova on the net of the cage and seventeen more next day, so we let her go and prepared for a big nursery.'

In their travels they suffered beds of coconut matting and

beds of bamboo poles, and pillows Miss Fountaine declared must be stuffed with cannon-balls; filthy and frequently limited food, and unfriendly villagers . . .

After a feast day when a quantity of raw spirits had been consumed and the excitement was at its height, I heard these people congregating outside the house, while a lot of men and boys shouted in no friendly manner 'Come on Americano! Come on Americano!' with occasional cries of 'Butterfly!' intermingled with the uproar. I was not in the least afraid for myself, as I knew they would not dare to hurt me, and this was all intended for Charles – Charles who only that morning had showered candy amongst some of those same boys whose treble voices now mingled with the coarser tones of the men in the darkness. My fear was that they should discover he was not here, but staying at another house, and go there. We were glad when these rejoicings were over. I did not know how bitter the hatred for the Americans is amongst these people; we lost no time in conveying the truth to them about our nationality, for as English we were looked upon with less distrust and dislike.

The Filipinoes are a queer people; not wanting in intelligence though by no means so clever as they imagine; those educated in the States appreciate the Americans but the people as a whole have but one idea, to be given independence, little knowing how incapable they would be of standing alone. They calculate that the awful earthquake which has just visited Japan would cripple that nation for many years and so do away with the menace of Japanese conquest. They miscalculate the determination of the Japs; the Filipinoes had better stick to their cockfights and not reckon on any victories where the Japanese are concerned.

Charles and I went to one of these cockfights, held in a squalid room in a back street of Los Banos. The noise is terrible; it is of course entirely a gambling proposition and the excitement of the people, together with the crowing of the cocks is deafening, so what with the stench of the audience neither ears nor nose has much to appreciate. It was just two wretched roosters pitted one against the other, with long steel blades supplementing the spurs nature bestowed on them. With these they soon inflict cruel wounds on each other until one runs to the edge of the arena and a heap of crumpled feathers is all that is left of him.

Christmas Eve was a very unhappy day for us both: we had returned to Manila and there is something in the heated atmosphere there which seems to have an effect on our nerves; we had a big quarrel, and I wandered about aimlessly for hours, too wretched to care what happened, until the daylight began to fail and it was soon dark. The Club was brilliantly lighted and I could see the swaying figures of men and women dancing, while little American children were going about with their parents in auto-mobiles, with the gladness of Christmas in their young eyes, while I felt like a vagabond. Why was I not like those well-dressed American women with happy children, or grandchildren as it would be by this time? I loved the wild life I had chosen, but sometimes I feel a longing for what my fate might have been otherwise, something so very different. I wandered slowly away and presently found my way back to the hotel, not daring to ask the bureau, as I passed in, if Mr Neimy had returned. It seemed a long time before his knock came at my door. Once reunited we were soon friends again, but 'to be wrath with one we love doth work like madness on the brain'.

For a collector like Miss Fountaine, rearing large numbers of caterpillars in order to get perfect specimens of newly emerging butterflies, there were conflicts between her need to keep on the move to find fresh species and collecting grounds, and the need of the 'nursery' to be kept supplied with the right plants. 'The country for miles around Hondagua was entirely devoted to the cultivation of coconuts, hills and plains alike, and a day spent a few stations further down the lines where there was plenty of glorious jungle produced no entomological results, neither were we able to find the milky shrub on which our larvae of the big *Euploea, E. Althaea*, were feeding.' Another move lost many more caterpillars of a butterfly, *P. Semperi*, 'for which Mr Longsdon, an English collector I was now corresponding with, was offering me £2 each for the female and 15 shillings for the male . . . the live female *Magellanus* we had brought with us hoping to obtain some ova of that rare and wonderfully beautiful species settled the matter by dying next day without having laid a single ovum.'

They had, however, captured the beautiful green *Papilio Daedalus* and the gorgeous *Appias Domitia*, and they would, if they could, have ended the season's collecting there. Miss Fountaine wanted to return to England where her sister Geraldine was making a recovery, it seemed, after being seriously ill with tuberculosis, and Khalil planned to visit his uncles in Chicago, where he had lived for some years in his youth and where there was a Greek Orthodox bishop through whom, they believed, he might at last get a divorce. But shortage of money meant they could not afford the fares. Miss Fountaine had sent instructions to London for the remainder of her War Loan to be sold, expecting to get between £250 and £300; it realised only £150, so they had to wait for her next letter of credit to arrive before they could pay off their debts and move on. To economize, they moved into one of the less desirable lodgings of their travels. There were no newspapers; 'we were buried alive just longing for our deliverance, we feel like prisoners' – but, she noted as her diary year ended on 15 April 1924, there was 'Charles' dear voice outside my door . . . he will try to snatch a good-morning kiss before I raise the little dimity blinds . . .'.

They spent their morning with their nets, searching rice-fields and strips of jungle, wading through sea-inlets or crossing creeks on fragile bamboo bridges, seeking the more valuable butterflies not only for their own collection but now for other collectors.

The thing they all seem to want most is *Ornithoptera Magellanus*, but this magnificent species is no longer here in evidence at all now; so that Mr Longsdon's offer of £7 for one male and two female will, I fear, never be met with. We always speak of this butterfly as '*Hondagua*' even when we are alone in the jungle, in order that the Filipinos should not know when we are talking of *Magellanus*, for people have been writing to them too, offering big prices for this much coveted treasure, so that they know it well by *name* though they have no idea what it is like or how to procure it themselves. One of their many mean traits is the fear that any foreigner should exploit their islands, and if they knew Charles and I were specially after this *Magellanus* I believe our lives would be in danger. Yet even if they had the slightest idea what it is like, they would be sure to confound it with *O. Rhadamanthus*, which also flies here, and much more commonly. So we are not taking the bread out of the mouth of any Filipino by trying to secure a few specimens.

She spent the afternoon attending to the net-covered cages of caterpillars and pupae, and dealing with their morning's catch.

It was past six o'clock before I was able to sit down and rest, taking a small peg of a kind of spirit they have here, made I believe from the sap of a palm; it is not very nice. After this I sat by one of the windows in my room and smoked cigarettes (a cigar is generally only my Sunday treat) while the darkness gathered quickly over the grassy space between this house and the old church over the way. The church is a most venerable pile indeed, fast crumbling to a ruin with its walls inside covered with mildew and outside clothed in a dense vegetation from which we have been told that an 'Ameri-cano' once discovered seven different species of snake. All good Catholics in Polillo still gather in full force every Sunday here to kneel in the dust of its unpaved interior.

I love to sit in the evening, when the day's work is done, often thinking of lands far away, places I have visited. And as I sit and think, through the puffs of my cigarette, I like to watch the children coming back, with their long bamboo canes full of water from a distant spring far away in the jungle; or maybe a buffalo goes slowly by dragging a slide, not unlike the slide we used to have for our horses in Australia. There is a bamboo seat round the trunk of a big acacia tree close by and here the girls of Polillo will gather as the darkness deepens (there is no spring twilight here), till by and by the youth of Polillo join them, and laughing and talking, thus spin out the simple story of their lives, after the manner of youths and maids all the world over.

Eventually Miss Fountaine's money arrived, and they set off, though by this time Khalil was resolved to return home to Damascus; a distant kinsman had become Bishop Neimy, Miss Fountaine reports, and Khalil would combine filial duty and his somewhat prolonged search for a dissolution of his marriage. So he sailed westward for the Mediterranean; she sailed on east, to Hong Kong, Shanghai, Japan and Vancouver.

At Hongkong more passengers came on board the *Empress of Asia* and I had two young women in my cabin, and we had lots of fun together, especially one evening when one of the stewards brought in a little note for me from some man who desired to know if I was a Miss M. E. Fountaine, who used to be with a company of stage actors and actresses who at one time were playing in Shanghai.

In spite of the laughing remonstrances of my cabin companions I insisted on merely writing 'Nothing doing', and returning the note by the steward. Miss Blyth afterwards went out and saw the man who was most anxious to know exactly what I was like, and all about me; poor devil, he was evidently terribly desirous of meeting that duplicate of mine again, and I began to feel sorry that my reply had been a little abrupt.

Shanghai was a horrible place, a great teeming Chinese town. It was here that I saw the truly pathetic sight of a little new-born baby girl being done to death. The wretched little creature was still blue, and could scarcely have been an hour old, but it was already strapped on to the back of a little girl, who was trudging pitilessly round and round a small courtyard, unmindful of the cries of the new-born infant on her back; she must go on trudging round and round that courtyard till those cries stopped, and she would then return with a dead child to its wretched heartless parents. . .

But Japan seemed quite different and the Inland Sea passing between islands very beautiful. So indeed was Nagasaki, an exquisite little place. . .

Miss Fountaine had booked second class for the Pacific crossing; she found the trip enjoyable but, in arranging the other sea-voyage part of her trip, from the East Coast back to England, had taken care to pay the difference and travel first class. The difference, inexplicable even in 1925 terms, was only £2. Between the Pacific and Atlantic voyages she had provided herself with sufficient time to spare during the crossing of the continent by train to spend a fortnight in the Rockies, at Banff; much of it was devoted to butterfly hunting, 'many happy hours with nothing to eat all day but delicious drinks from mountain streams'. Finally she had to return to England – and the Customs.

I had brought a lot of Presidente cigars with me from Manila, which had come across Canada in bond, but I was a little bit anxious about the smuggling of them into England. However, the man who examined the luggage at Liverpool only murmured something about cigars as presents, and then became interested in a few cigarettes which I was conscientiously owning up to, having really supplied myself with them to take him off the scent of the cigars. And all the while on the other side I was being pestered with the reporter of a Liverpool paper, though why he failed to find any more interesting passenger than myself I can't think! I reached Euston at about 11 pm having come a journey of over 10,000 miles, and crossed the two big oceans without a hitch and with sixty excellent cigars worth in England about two shillings each, for which I had paid in Manila about one-and-a-quarter pence.

In London I was extremely busy labelling all my specimens and packing up butterflies for sale, some going to America, and others to various collectors in England. Mr Longsdon had been to see me. My first impression, when Mrs Bird showed him into my studio, was a shock: he was not only quite abnormally ugly, but so queer in his manner, he spoke all the time with his head turned away and eyes averted . . . a most uncomfortable impression. When I accepted his invitation to go and see his collection of *Papilioniae* at the Flower House, Beckenham Lane, I don't quite know why, but I said my brother-in-law would probably be with me (I was expecting Hill and Geraldine to come up to town). However, on the day fixed for my visit, Hill and Geraldine had not arrived. My strange host met me at the station, and seemed quite taken aback to find Hill was not with me.

'But how brave of you to have come alone!' he said.

I found the Flower House and its environs quite astonishingly beautiful, a large and beautifully laid out landscape garden, filled with the most gorgeous flowers of which a man we met there insisted upon gathering an enormous bouquet for me. The house itself was very large and the room upstairs where Mr Longsdon had his collection was well appointed and very big; his collection too was very fine, though he said he had not caught any of them himself, but had purchased [them] at an expenditure of about £100 per annum. We went down to another room for tea, a most sumptuous repast, and again a beautifully furnished room with every luxury; but I was beginning to feel a curious strain, and the longer I stayed in this house the more uncanny I felt it to be, and yet I did not feel as though it were haunted. It was something of a relief when I, with my big bunch of dahlias, went back to the station. But I had not nearly finished seeing all Mr Longsdon's collection, and it was arranged that I should come again, and this next time I decided it would be with Hill and Geraldine.

Mr Longsdon had told me that he had at one time been an artist, and had exhibited in the Royal Academy. Why he had given up art for entomology was one of the mysteries which surrounded him. Naturally Geraldine was anxious to see his paintings, and after considerable pressure he began to show them to us. Those paintings were indeed the work of an artist, but so infinitely sad; what tragedy could there have been in the life of this man to make him paint like that? When we began to express our delight in what he was showing us the poor thing got so pleased and rubbed his hands with pleasurable satisfaction; but we had unknowingly touched the wrong chord – he was never sane again afterwards, and gradually it dawned upon us (we had seen one or two queer-looking people walking about, and had heard some disconnected playing on a piano in the next room, while we were at tea) that this was nothing but a private lunatic asylum, and this poor thing with his kind gentle disposition one of its inmates. 'Not a very bad case,' Hill said afterwards. But he was bad enough at the end of that day.

Charles, in the meantime, seemed to be getting on very well with his father's cousin, the Bishop, and by and by he wrote to say that this worthy prelate had dissolved his former marriage, making him a free man. But, of course, Charles had to stay on in that place, on account of his mother; he would, I knew, not think of leaving till the tourist season was well over, so as to earn all the money he could for her; which was perhaps as well, as my income being so crippled, what with the iniquitous income tax and one thing and another, there was a limit to its capabilities. However, I was selling butterflies at a great rate, sending off packages, and having men come to see me; I disposed of two females of *Papilio Semperi* at thirty shillings each.

Florence Curtois [a girlhood friend] came to tea, and a more appreciative admirer of my butterflies I could scarcely have had. Lisle [her brother] too came to see me, the disconsolate widower one minute and the next admitting himself to be in love with a young girl of eighteen. I did my best to be sympathetic on both topics; also about his book, which I believe he has been writing for years, but if it is calculated to bore the general public as much as he himself invariably bores me, I'm afraid there won't be a very great sale for it. Then he announced himself to be a communist, and expressed the utmost disgust at the results of the recent General Election, and its sweeping Conservative majority; in which sentiment I failed to be sympathetic. However, when I went to lunch with him in his hotel a few days later I liked him much better.

The next day after that Lord Rothschild had invited me to spend the day at Tring, in order to see his wonderful museum. I lunched with them, en famille, which consisted of the Dowager Lady Rothschild and one or two nondescript ladies, who did not contribute much to the general flow of conversation. This was mostly kept up by a very young girl, daughter of the late N. C. Rothschild, who had been so heavily weighed down with the responsibility of his immense wealth that suicide had appealed to him as the only way of escape; but his daughter was a delightful child, barely grown up, and full of life and the joy of it; she loved to recount her exploits on the hunting field, much to the anxious concern of her stately grandmother. In spite of this child, there was all the cold grandeur of a country house in England, with all the evidences of great wealth. What a contrast to Lisle and his communism! And yet I never felt a speck of envy for any of them. Of course, I sat next at the table to Lord Rothschild (who the young girl addressed as Uncle Walter) and I did my best to make conversation, in which I fear I only partially succeeded.

As to the Tring museum, it was infinitely more wonderful than anything I had ever imagined. I felt inclined to quote the Queen of Sheba's remark to Solomon. 'The half of it had not been told me!' But I was not quite sure that the Jews might not hold Solomon and possibly also the Queen of Sheba in such strict veneration that the levity of the observation would be considered inappropriate. It was not only the insects that were so amazing, every branch of natural science was represented with practically every known species, and all in such perfect order as to be almost life-like. As to the lepidoptera, Lord Rothschild told me that the number of specimens he now had of butterflies and moths alone was more than one and three quarter millions, and I thought to myself how my poor little collection of some 16,000 butterflies would be merely a drop in the ocean.*

The winter in England was coming on apace now, and the short days

---

* Half a century later I wrote to that 'very young girl', Miriam Rothschild, now herself a distinguished naturalist (and author of a book on gardening to encourage butterflies). 'Alas,' she replied, 'I can't help you much. I was 14; she seemed a very ancient lady to me. All I put in my diary was that she kept her end up with Uncle Walter very well where Latin names of butterflies were concerned. . . .'

and the cold, filthy black fogs of London were having a most depressing effect on me. I left the dismal old city in a wet fog, nor could anything have been more dismal than the North of France. I passed the night with three old men in the carriage with me, one an Indian veteran, a nice, friendly old thing; another who was guarded and unfriendly as though he apprehended that I might be inclined to take liberties; and the third was a Frenchman who snored so loudly and so persistently that I asked the Indian veteran if he did not think I might give the snorer a discreet kick, to which idea he most readily acquiesced, so I did it, with results most satisfactory. There was absolute silence for about five minutes, and then the infernal row began again as bad as ever.

Next morning we found the sun shining brightly, and Hyeres bathed in sunshine and quite warm. For me at least the agonies of an English winter were now over. It was also quite a relief to dispense with all the discomforts and inconveniences of Lexham Gardens, and instead of waking up on a bitterly cold morning and having to boil my own water over a spirit lamp to wash in, to wake instead in a warm, central-heated room with hot and cold water 'courante', and just lie and wait till the femme de chambre brings in my coffee and rolls. Of course I was lonely here, but Charles sent me a special letter with good wishes for the New Year, saying that he hoped 'the next New Year we will be together forever'.

I was busy with my article on collecting in the Philippines which occupied most of my time in the morning; in the afternoon I would go for long, lonely walks over the hills round Hyeres hunting for *Jasius* larvae, quite scarce now, to send to America.

> However, weeks spent alone, and delays in Khalil's letters,
> drove Miss Fountaine again into depression and self-pity . . .

I often longed that I might die, and be put out my misery. No one would mourn for me if I did; my sisters I think in common decency keep it to themselves but inwardly would feel complete satisfaction at having a nice little addition to their incomes; my brother Arthur would, of course, be entirely of the same opinion; and as to Charles, I felt now he did not care a damn about me; in fact I did not believe he cared for anything except his old mother, that old hypocrite who will, I suppose, die quite happy some day, feeling no remorse whatever at having ruined both our lives. I wrote Charles a letter which I meant to be practically of the nature of an ultimatum:

'So long as your Mother is well and happy, you care very little, if at all, what becomes of me, otherwise you could not go on staying away from me like this month after month,' I wrote. 'If that is so I will only ask you just to let me know at once, (and I will) go away to Brazil by myself. I think this is

80

what you would like to happen, and then you can stay on with your Mother for the rest of your life. . .'

When I posted this letter at the Hyeres post office, next moment I was longing to be able to recall it. I had not been justified in writing to him like this, merely because he was staying on to make a little money for his old mother. I was to blame to a very great extent, and I knew it. And when the answer to that fatal letter came, Charles had literally taken me at my word – a thing they say no one ever wishes to happen, more especially a woman. 'I am very sorry to hear you are not receiving any letters from me, and you think I do not care of you any mor, my dearest you making big mistake', he wrote, 'I am no doubt in great fix between you and my old Mother I do not know what to do, I cannot hert the feeling of my poor old mother and leave here which she have nobody look after her or give her lofe of bread and she is very happy I am with her, and I certen if I go away she will die and I will have no hapeness or good future will be for me. God he will never forgive me. I do not like to be away from you and here I am in great fix betwen you and poor old mother I do not know what to do. . .'

Miss Fountaine, trying to justify the mysterious ways of Providence, could only conclude that her unhappy position was a judgement on her: 'Many years ago, when my own poor mother used to implore me not to leave her I used to boast that her entreaties had never once retarded my departure.'

However much Miss Fountaine might protest, and grow angry, and repent, and grow angry again ('Khalil's mother was in a rather pitiful condition, though she has been running this dying stunt now for four or five years. Her age is about 81, but many old women live much longer, his grandmother reached 95. . . .'), all was in vain; Khalil could not, would not, leave Damascus until his mother could be cared for. They had in the past discussed the old lady's living with them after their marriage, a prospect one can only view with wild surmise; it does not seem to have been raised again; the old lady did not want to leave her native land and Miss Fountaine does not appear to have considered going there; she had exhausted its entomological possibilities long ago. Khalil continued in Damascus, putting by money from his work as a guide and writing affectionate letters to Miss Fountaine; Miss Fountaine continued in London visiting friends and being visited, putting her collection to rights, going to the Derby (and winning 35s for 5s on 'Manna'), and looking for a new studio in place of the one in Lexham Gardens, which wasn't big enough to take a piano. It was a long search.

What queer places, and what queer people did I come across in my various quests for suitable quarters. Once it was a thin slip of a room apparently a repository for sponges; sponges (rather small ones) were there in thousands, piled right up from floor to ceiling, leaving only just standing room for me and the young lady who desired to let it. Then a very large room in a Ladies' Club, really most desirable in many ways, was not to be thought of, for the regulations forbade any individual of the male sex so much as to cross its sacred threshold, except on Wednesdays and Sundays; so that deal was off too.

She interrupted her search to join a party of women listening to a debate in the House of Commons: 'The conclusion I came to was that however much they muddled matters in their attempts to govern this mighty empire, they managed to do themselves extremely well. Baldwin, the Prime Minister, was there, with Winston Churchill sitting next him, and both gentlemen were lolling with their feet on the table; except only when the latter got up to make a speech, so badly articulated that it was scarcely possible to hear what he was saying. The best speech was made by J. H. Thomas, a Labour Member; and Ramsay MacDonald, with his tall figure and thick bushy grey hair, was physically by far the finest specimen of a man amongst the lot of them; but he was quite mad, never still for a moment, wandering about all the time, now whispering into the ear of an obviously bored usher, and then passing on to another.'

She at last found a large studio at 100A Fellows Road, Hampstead, her home whenever she was in England for the rest of her life. In it she set out her big butterfly cabinets across the centre to divide the large room; on one side her workroom looked out across the garden, on the other was her piano, a small table for tea when visitors came, and other furniture: 'her boudoir', she called it. She took her meals and slept at a boarding house in Finchley Road, not far away.

Not long after she had moved in, the owners of the house in whose garden the studio stood decided they wanted to sell; this meant, Miss Fountaine realized, that her lease might not be renewed when it expired the following year – doubly inconvenient because she was planning to be in West Africa on another collecting trip by that time. It became obvious to her that she should herself buy the whole property. Cropper, her accountant and adviser, was opposed to the purchase, perhaps because it was a leasehold property, 'but when I calculated that I should have to live to over 102 before the lease expired, and the ground rent was only £10 per annum . . . I wrote to Cropper and told him I wished and intended to purchase the whole property.' She succeeded in getting the lot for £2400, and found herself now the landlord of the man (a naval commander named Smith-Wright) whose tenant she had briefly been.

She had left this business unfinished, to be concluded for her by the ever-reliable Cropper, when she set off for West Africa; she had also left behind Geraldine, now very ill with consumption. They had spent a great deal of time together in the last weeks before Miss Fountaine sailed: 'We would wander together through the far-distant days of childhood, only I could sometimes wish poor Geraldine did not always dwell upon the sorrows and cruelties we had to put up with when there was so much that had been very happy too. I used to sit for hours sewing away at a colossal piece of needlework to make a covering for the backs of the cabinets in my studio, which certainly did not make a very pretty barrier from the boudoir side . . .'

Geraldine's illness worried her, and not only for Geraldine's sake. 'Must not Evelyn and I have the taint of this horrible disease in our blood? And if we are past the age for developing it, why had poor Geraldine fallen a victim to it in the latter fifties?' She did not expect to see Geraldine alive again when she sailed from Southampton for the Canary Islands *en route* to West Africa.

As soon as we got clear of the Bay of Biscay things assumed a more favourable aspect, and everybody began to come up like so many rabbits out of their holes. Mr Pomery, a government entomologist from Nigeria, gave me information, while I on my part undertook during my stay in the Canary Islands to try and breed sandflies (*Simulium*) for him and preserve specimens of them in their various stages; he described how I was to procure them, their larval and pupal states being passed in running water. The work would be something quite new to me, and though not without considerable difficulties, I was quite looking forward to the task.

On Tenerife one day I found myself in a Hudson motor-car in company with eight young Spaniards, flying at headlong speed over the mountains back to Santa Cruz; one of them spoke English and was most friendly; and what a relief it was to know that his attentions were merely prompted by a desire to impart knowledge, and to show off his English, which was really quite passable; and when he chanced to remark conversationally that he was a 'naturalist', I became quite interested, seeing at once the possibility of obtaining some information that might help with the elucidation of the

Simulium problem, but he quickly corrected himself, saying he had meant to say 'vegetarian', whereupon my interest flagged. However, he continued to keep me posted up with any information he thought might be of interest to me, while we sped along at a furious rate, racing every other car on the road to the imminent peril of pedestrians, various donkeys, mules and bullocks, not to mention an occasional camel, which we encountered on the way. Was I frightened? the young man enquired. No, indeed, I could most truthfully assure him that I was not, though it was rather a unique position for an old woman over 60 to be in, rushing madly over a mountain pass in a car packed to overflowing with young men, and I could not help thinking it would not be altogether pleasant to die messed up with all those Spaniards!

> When she did arrive in West Africa she thought for a while she had taken on too much.

The heat in Lagos was really terrific, and there was no cool hour before the dawn when, in the shades of my still dark bedroom, two dusky figures would creep noiselessly in, with my *chota hazri* – in African language 'small chop'. Lagos was surrounded with mangrove swamps, enveloped day and night with a heavy, stagnant heat, and at first I did feel that I had this time struck a climate which I should not be able to stand. On the days I went out for a few hours I certainly felt the heat very much less, but unfortunately there was little doing amongst the butterflies round Lagos. Mr Wakeman let me have the mission motor car at the very moderate rate of sixpence per mile, provided it was not otherwise engaged. I went six miles out, and did fairly well in light jungle, finding the natives quite harmless and most civil in their friendly greetings, as I went past. But these expeditions by car soon came to an end, for two of the missionaries took it for a month's trip up-country. The two most essential items for a woman in this country are a car and a husband; the one who should have been my husband was far away, and the car alone did not appeal to me as a very desirable acquisition, in the extremely dependent position I was in.

> In spite of these difficulties she brought her usual energy to her work; on 15 April 1926 she hired the mission car again, was driven to a bush village and then walked on alone, net in hand – to spend five hours losing herself on paths 'all seeming to lead me deeper and deeper into that dense jungle'.

I was beginning to despair when I saw, coming towards me, a very tall man, dressed in an ample, clean white burnous, and carrying a large light-coloured umbrella to shield his huge person from the scorching sun; he looked like a Moslem of a better class and smiled pleasantly when I accosted him. Though he did not seem to know any English, I managed to convey to him that I was desirous of finding myself on the big road to Lagos, while he on his part managed to convey to me the fact that I was then going exactly in the wrong direction, and that I had better follow him. So with this big Mohammedan striding on in front I just pattered along in his wake, while from time to time, at what he would seem to consider any unnecessary lagging on my part (for instance if I paused for a few seconds to catch a butterfly), he would look round, and beckon me to 'come on', and I think those two words were about the extent of his English vocabulary.

Once we passed a great, dark pool in the very heart of the jungle, and I thought of that very apt description I had once read of such pools as 'like the blood of the trees'. However, after long wandering and many fearsome doubts – I am rather ashamed to confess I mistrusted the man – we emerged at last to where cassava was growing; we were getting back to civilization.

Her guide had brought her not only to the main road but to a friendly Salvation Army mission – 'Oh, that cup of tea! Shall I ever forget it!' And she still had energy enough, stopping on the way home to pick food for her caterpillars and to run after the watching children and pretend to catch them in her butterfly net, 'which caused an immense amount of amusement'.

A few days later she had booked a short passage aboard a German cargo vessel to Victoria in the Cameroons. Miss Fountaine was still uneasy about the Germans; from being rather pro-German in her youth she had made the painful change demanded by wartime propaganda, and now found it difficult to cease regarding them as depraved – but her German was good and the ship's officers hospitable.

There is pasted into the diaries a photograph of Miss Fountaine in the Germanic solidity of the *Irmgard*'s officers' mess. The perpetual sweat of West Africa gleams on the men's faces, their white jackets are crumpled in the heat, the electric fan blurs as its blades push the heavy air, but there is a solid German meal in front of them, and a bottle of wine. She is seated next to the youngest of the group, the ship's doctor; she looks coolest of the group and – perhaps it is a trick of the light – just a trifle like the cat who has got at the cream.

NINE

Love and Change
1927

*The young German ship's doctor ; Germans as colonists
—— elephants and fever —— falling in love with Herr
Rein —— but faithful to Khalil —— the French as
Pagans ; and colonists —— Khalil's letters, unchanging
bad spelling in a changing world —— a little stout
gentleman with gold teeth, more son than lover —— to
Guadaloupe and Martinique —— hotel walls so thin
when she taps cigarette man in next room cries Enter !
problem with a sermon : what did the Lord do for
clothes ? —— Charles ill with fever —— Nothing could
help or comfort her now*

I soon became aware that a young German with a boyish face and rather wistful blue eyes was showing signs of a desire to be friendly with me. He was the ship's doctor, and, I suppose, because there was no other white woman on board, he suggested that we should go on shore together. He said his English was 'small, small', but as there was always my German to fall back upon we got on quite amicably. My inclinations were to get out as far into the bush as possible, and his leanings were towards the villages to get studies of native life for his camera, so we were in a chronic condition of losing each other. This most amiable individual told me he was 27, and that at 17 he had been fighting in the German army in France till the end of the War; but, like all the other Germans I met, he seemed to feel no resentment. I wonder if they are really as amiably disposed towards us, as they seem to take so much pains to impress us with.

Our next day on shore was spent at Degama, up the Bonny River; and it was here that I discovered that Dr Deussen's favourite models were little naked Dinahs; and on one occasion when a group of natives were posing for his benefit, he confided aside to me that what he *would* like would be if the big ones would also disrobe in favour of his camera, and did I think, if he 'dashed' their husband's sufficiently that he might suggest it? I vetoed the suggestion instantly, saying that these men are often very particular about their women and would be very much hurt, and somewhat regretfully he abandoned the idea.

Ashore in Victoria I engaged a cook. His credentials stated he was a good cook, and clean in his work; I afterwards discovered they were a fake, but I had no cause to regret engaging him – at £2 per month, he providing his own chop. I left my new servant to watch over my luggage on the little wharf while I went off to see the Assistant District Officer, Mr Stone, a nice clean young Englishman, who at once arranged that I should be put up in the Rest House. The Rest House was roughly but sufficiently furnished; however, cooking utensils had to be provided by me, so, having collected my cook and the luggage I set out to find these adjuncts; some I purchased at the African and Eastern Trading Company, others off stalls, which lined one of the principal thoroughfares, while the cook followed me with a big basket to carry my various purchases, crockery, cutlery, cooking utensils and provisions; so that I eventually dined that evening in the Rest House, quite sufficiently if not sumptuously.

The next day I engaged a 'small boy', a very solemn looking child, who was recommended by the cook; and then I set to work to explore the bush, the small boy always accompanying me and carrying some of the necessary paraphernalia, and I was soon too busy to be lonely. Later, I went up to Buea, where I hired a furnished bungalow at £6 per month. There were many bungalows at Buea, most now standing empty, each one being provided with a bathroom, with water laid on, and also perfectly arranged sanitation. Buea was beautiful, with its miles of hedges all of roses perennially in bloom, everything having been fixed up by the Germans before the war. I already began to be astonished at the colonising capabilities of these people, which hitherto I had been led to believe inferior to our own.

The collecting was very good, and the climate, considering the elevation, still bearable, though colder than I cared for at night, and too much inclined to rain during the day, as it was now the end of May, and the wet season was coming on apace. The Residency up on the high ground was really a magnificent building, formerly occupied by the German Governor of all the Cameroons, he having, I was afterward told, spent two million marks sent out by the German Government for the development of the country, entirely on the erection and beautifying of his own residence, which had naturally created a big scandal. Now this palatial building was occupied by our Senior Resident, who came to call on me and invited me to dine at the Residency.

As the summer advanced the climate of Buea did not improve, and I was

told that at Ekona, some twelve miles away and at a lower elevation, butterflies abounded. It was the centre of a big cocoa growing district, now once
more in the hands of a German company; and all the houses there were
occupied by Germans, the principal dwelling being the house of Herr Rein,
the manager of the whole concern.

I wrote to Herr Rein in German, and got a reply that I might stay for
one or two weeks – though in the meantime I heard that Herr Rein had been
saying that, as a single man, he had misgivings about having a single woman
to stay fearing she might want to marry him – besides, what would his
employees think? I hastened to have it conveyed to my future host that I
was not by any means a young woman, and he could set his mind at rest.
So I went to Ekona, sending all my luggage on a lorry with the cook to look
after it, while I myself prepared to walk down with my net and small boy,
rather to the disgust of the latter, who would have much preferred to have
gone with cook on the lorry. Before we reached Ekona, very heavy rain came
on, so that I and the small boy (whose only shelter was a banana leaf) got
soaking wet. But the sun came out again and already I saw and caught some
butterflies I had never seen before.

Herr Rein was anything but typically German in his appearance; a spare
man, rather below the middle height but with that air of distinction that is
the outcome of good breeding, very dark, with a small, black moustache which
did little to hide an exceptionally ugly mouth, the ugliness of which however
was entirely redeemed by an exceptionally charming smile. In that first
moment of meeting the impression I received was that he was agreeably
surprised; he looked his approval, and like a flash it was gone. I only recalled
that look, amongst other things, some weeks later.

I stayed as his guest, dining with him every evening. Every day I found
species new to me, and often saw rare females laying ova. Herr Rein had
given me a native from the cocoa estate as guide, and my small boy used to
come too, and was getting quite good at collecting specimens. He got terribly
scared when a herd of elephants turned up in this neighbourhood, especially
one day when a sound in the forest up the steep incline right above us as of
great beasts breathing even gave me a scare, though of course, to allay the
fears of the natives, I attributed this ominous sound to anything in the world
other than elephants. But elephants it certainly was; that night at dinner,
after the servants had left the dining-room, I told Herr Rein about it, and
he instantly imitated the sound exactly, assuring me at the same time that
there was no danger of them attacking.

Next day the small boy started a bad foot, which only recovered after I
had repeatedly assured him that elephants did *not* eat small boys, and if we

did come across them he could climb up the nearest tree, and leave me to settle matters with the elephants.

There is one butterfly on the West Coast of Africa which for some years has been known amongst entomologists as 'the elusive *Drurya*', which, though occasionally seen at various places by several people, has up to the present always managed to evade capture; it has been described as being of a wonderful vivid blue, and very large, and it was now my keen ambition to be the first to catch a specimen of this gorgeous creature. One evening in Ekona I had, I thought, seen it down by the stream. The lovely creature was sitting with closed wings, imbibing as one might say a last drink before retiring for the night. Just as I was making up my mind to plunge across the stream to it, the thing got up and floated majestically over our heads. It was of a most beautiful sky blue, and jumping at a conclusion I cried, '*That* is the butterfly I have come to West Africa specially to take!' making at the same time a frantic effort with my big, yellow net to catch it, but it was just out of reach, and soaring high into the air it sailed proudly away towards a patch of thick jungle.

One morning, only a few days before I was to leave, I had got up as usual feeling quite well when, with a suddenness which was in itself alarming, I felt as if I were dying. I lost all power and sense of feeling, while a horrible agony was creeping all over me; I felt literally paralysed from head to foot. I could hardly move, but by holding on to chairs and leaning against the wall I got back into my room, and creeping in under the mosquito net, lay flat and almost lifeless on my bed. For over an hour I must have lain there, till with an immense effort I managed to get up and get at a small bottle of brandy I had provided myself with, and after drinking that I felt slightly better. The cook showed real concern when he found me thus, and brought me in some tea, and then I began to feel much better, so that I not only dressed but afterwards went out that morning as usual. Another attack came on in the bush, when I was far from home, but it was not so bad as the first. But all through dinner that evening I felt the beginning of fever. (It has since occurred to me that this must have been a slight stroke of paralysis, no doubt subsequently cured by my having malaria.) I lay in bed that night with my head and face burning while my body shivered horribly, and my one thought was that this would be the end of my collecting in the Cameroons, but next morning I woke up in a tremendous perspiration, and after that I felt quite well.

Miss Fountaine moved on to Meanja, travelling some miles on a monorail trolley, varied by a ride on a small engine which 'jumped the rails and dashed into a wall of rock, I in the meantime promptly jumping off on the other side . . . but no one was any the worse'. This after a bout of fever and the mysterious attack of paralysis – and she was now well into her sixties. . .

Mr Rein rode down to Meanja one day and I met him in the forest on his way back. How charming he was, dismounting at once from his horse to stand and talk to me; though I begged him not to bother to do so; and he just acquiesced at once, with that wonderful smile, to my suggestion that I might return for one week more to Ekona. I noticed how the flies had bitten the ears of his horse, rather a fiery little chestnut who seemed rather impatient that his master should dismount here and hold a conversation, however short, with someone in an old dilapidated topee, rather soiled skirt, and a very disreputable pair of gloves.

The night before I had arranged to leave I had another attack of malaria. I had intended to trek back on foot, but when Herr Luce, who had come to superintend the packing of my luggage on to the trolley, saw me stagger and nearly fall from weakness, he sent for the passenger trolley. A monorail trolley is not a luxurious mode of travelling, jolting and swaying along, every movement causing me to feel more deadly ill. There were several delays too, where the line was up. Rain had begun and I became so dreadfully ill I had to get off and lie down on the rain-soaked rocks at the side of the track. My own servants were in advance with the luggage trolley, so there I was, quite alone, with wild African bushmen who just sat and watched me, unable to do a thing to help. I thought of my poor mother, who was so nervous that she would not go alone a few doors up the street from her house in Bath to see her friend without having Lucy or one of us with her, for fear of being taken ill!

A few hours later, at Herr Rein's bungalow, she was recovered; malarial attacks are brief. There are other fevers less easily overcome. When she came to leave at the end of a week Herr Rein persisted in inviting her to Sunday lunch; he seemed irritated when she refused and reminded him that she never took lunch. It was not until she was back among the urban delights of the little town of Victoria a few days later – 'bridge evenings with "small chop" and other little social functions which I much enjoyed' – that she saw Herr Rein in a passing car and realised, 'like a flash of lightning across the calm serenity of a summer sky', that she had fallen in love again.

I was mad! I loved that man, and I always had loved him all the time. But why the embers of a passion I had never owned to myself for so much as one moment before had thus suddenly burst into flames I could not imagine. The air was singing with a thousand rapturous voices, while every tree and shrub, every flower and leaf, were glistening with rainbow coloured hues; for that intoxicating glamour which is solely the heritage of youth was with me now. For several days I lived in that mad dream of happiness which lends a radiance scarcely of earth to everything around one. I remembered the day we had met on the forest path, near Meanja, when he had got off his horse to speak to me; and I saw again that fiery little chestnut so impatient at not being allowed to carry his master straight back to Ekona, and how many other little episodes and incidents did I now conjure up, especially that last day when he had begged me so much to take lunch with him. Why had he been so anxious for one more tete-a-tete meal with me?

I tried to reason with myself, recalling that my advanced years must stand as a barrier between all possibility of this passion being returned; but then no one thinks me anything like as old as I am – even Lisle Curtois, barely two years ago, had looked at me and said: 'Why, Margaret, to look at you, you might still be in the forties!' And Charles, my own dear Charles, how often has he said: 'You have an attraction, no young girl, she has it!' But I was a fool.

> And, not without a struggle, Miss Fountaine put Herr Rein out of her heart, if not completely out of her mind. For one thing, there was a new batch of letters from dear Charles to read, and his faithfulness pricked her conscience; 'a woman who could not be true to such a lover was indeed worthless'. She hunted butterflies around Lagos and Freetown, and when she did stop for a while at Conakry in French West Africa she was so depressed by inactivity that she travelled four hundred miles inland lest she should 'go melancholy mad'.
>
> She spent Christmas 1926 in French Guinea, noting that 'the French bourgeoisie celebrated the occasion by making a deafening noise singing ribald songs, most of them becoming so drunk it was difficult to say what lengths they would not go to'. But she found an interesting *Charaxes* larva . . .
>
> The eleventh volume of the diaries, covering the six years to 1933, begins with Miss Fountaine returning from West Africa to the Canary Islands, where she spent her 1927 'day'. She had received letters regularly from Khalil, and now wrote sending him money to pay for someone to care for his aged mother, and for his fare to Marseilles, where they would meet. She had little money left after her travels, but Khalil was worse off; it had been a year of riots and banditry in Syria, frightening away the tourists on whom his living depended (though by this time he had, it seems, a sideline in preparing and selling a hair-tonic; Miss

Fountaine thought it excellent). Khalil wrote that, though he disliked being so poor man, he was coming, and would never leave her alon any mor; his spelling was an unchanging thing in a changing world. While awaiting her ship she played her own touchingly ridiculous version of Patience, 'and was pleased when the Queen of Hearts (myself) joined the Knave in the completed sequence'. The Knave was Khalil, of course. 'How foolish we are sometimes, and what small matters occupy our poor silly brains!'

When she sailed for France, with two young Frenchmen among her fellow-passengers, she reflected how she always got on much better with foreigners than with Englishmen . . . 'no doubt the latter have their points, but with my own countrymen rarely if ever have I felt that magnetism and mental affinity which draws me to men of any alien race'.

In Marseilles there was no Khalil – only more letters, one assuring her that 'we fly away to some new country, we are two wild birds'. She had in fact thought of flying literally; of 're-turning to London by "avion", which would have meant about a twelve-hour flight, with stops at Lyons and Paris', but problems with her luggage made her abandon the idea. It was just as well, she decided when Charles at last arrived, because he would have vetoed it. Charles had indeed arrived, sporting a new set of front teeth all of pure gold, which so much impressed the young lady receptionist at their hotel that she put him down at once as a New Yorker. Miss Fountaine wished he had had them white, but to worry over so small a thing when she had him safely with her again would have been a sin, she declared.

Though she thought of Charles now 'more as a son than as a lover', he never the less tried to make arrangements for them to be married by the British Consul in Marseilles. This function-ary, apart from annoying the would-be bride by behaving as though she were not concerned in the matter, raised legal prob-lems about Khalil's British citizenship; he was undoubtedly British in Australia, and possibly Australian elsewhere; Miss Fountaine would have his nationality upon marriage, whatever that might be, losing her own. Miss Fountaine, prophetic of a feminist controversy a lifetime later, wanted to know how the consul would feel if, on marrying a French girl, he found himself a French citizen without the option. It was clear, however, that she could only be certain of retaining the nationality to which she attached so much importance while still marrying Khalil, if he were to spend an unbroken year in Britain and then take out naturalization papers. To Britain they went – by train and boat, not 'avion'.

In London they stayed at separate boarding-houses, spending each day together at her studio, he reading the papers and writing letters while she unpacked and labelled the spoils of the last expedition. They made sightseeing tours about London; she was also kept busy helping clear up the affairs of her dead sister Geraldine as well as those of Geraldine's husband Hill, who had survived his wife by little more than a year. In spite of this sad task, Miss Fountaine was happy, even though 'poor dear Charles is no longer very good-looking, having become decidedly portly in his middle age . . . people must think me positively cruel, the way I make this little stout gentleman run . . .'. But the weather was very cold as the year advanced, and with November came an offer from the wealthy though lunatic lepidopterist, Mr Longsdon, to pay her return fare first-class to the West Indies if she would collect for him there. It meant six months in the sun; Charles could follow when his qualifying year in England was up; she could add to her own collection. With Charles's approval, she accepted; the parting was painful but 'Charles was still smiling and showing his gold teeth' as her train pulled out. Even so, it was a parting she was to regret bitterly.

Her commission from Mr Longsdon was to collect the various forms of *Papilio Polydamus* in the islands of the Lesser Antilles, the British St Lucia, Antigua, Dominica and St Vincent and the French Islands of Martinique and Guadaloupe. The required butterflies were hard to locate and to capture – one of Mr Longsdon's eccentricities was to require caught and not reared specimens – and while some of the islands themselves were beautiful and the weather warm, getting about was difficult and expensive. On St Lucia, for example, hiring a car was costly and the bus attended with dangers. On the precipitous roads travellers were 'at the mercy of a coloured chauffeur who did run the cumbersome vehicle over the side of the road on one occasion, when one foot further and we would all have gone rolling down the side in one horrible confusion of bodies, boxes, yams and other vegetables'. She disliked being tied to an agreement so limiting to her freedom, and was further irritated by letters from Mr Longsdon giving her ever more and more detailed impractical instructions; he was, she wrote, a lunatic, 'though nothing to the lunatic I was to have undertaken such a job'.

Guadaloupe and Martinique were lovely and the collecting was improving, but the little towns were 'wretched'; 'squalid';

93

'an absolute pesthouse of stinking filth and disease; no wonder typhoid fever was there . . . the combination of French and Negro being far from elevating spiritually, mentally or morally,' Miss Fountaine roundly declared. The locals cheated visitors and the hotels were generally bad – one was 'a filthy place'; another, because the proprietors had quarrelled with the local electricity company, was without electricity or water; a third had partitions between rooms so thin that 'once when I tapped my cigarette prior to lighting it a Frenchman next door called out "Entrez!"'

On the other hand, she met for the first time able and intelligent Negro office-holders, such as the director of a botanical gardens, 'a man of considerable ability', something she had not noticed on the English islands. She does not appear to relate this to the different racial and colonial attitudes of the French and English; she was shocked to find the French on Martinique 'of a low class and apparently for the most part of bestial habits . . . I was told it was no uncommon thing for a planter out in the country to have in addition to his French wife a Negress whose children would be brought up indiscriminately with the children of his lawful wife . . . what filthy swine some men are . . . it was good to be back in Barbados at the Balmoral Hotel.'

By the time her 'day' came round again in 1928, she had completed nearly all the commissions given her by Mr Longsdon, had secured a number of specimens for herself, and was preparing to return to England. But not to Khalil; the cold of an English winter – and the wet; this was the year when low-lying areas of London were flooded by the Thames – had proved too much for him and he had written to say he was accepting an invitation to spend three months in Egypt with a man he had met in a City bar. Miss Fountaine was disapproving – he was losing his chance of naturalisation – but blamed herself for accepting the West Indies commission and so leaving him alone. She was more perturbed when she found he was now back in Damascus. Before her boat sailed from Kingstown on the island of St Vincent, she attended church and prayed for Khalil; it helped and comforted her, though the service raised other problems.

The sermon was preached by the Bishop; being the Sunday after Easter it was about the risen Christ, and though he took for his text John XX 7, 'and the napkin that was about his head not lying with the linen clothes but wrapt together in a place by itself', he did not throw any light on one point which

I have often puzzled over and found rather disconcerting.

> This delicate point untouched by the Bishop was, Miss Fountaine explains, what the risen Christ did for clothes, the grave-wrappings being left behind in the tomb and His own clothes having been taken by the Roman soldiers. It was puerile and childish, she admits, but she wished the Bishop had suggested what seemed to her the only possible solution: the angel had brought fresh clothes with him. But that was taking a narrow material view . . .

The diary for the year to 15 April 1929 begins:

How can I bear to tell it? I dread to tell it, as one who with an unhealed wound would dread to see a rough hand about to be pressed over it. First the home-coming, which I had so much counted on, always thinking how delightful it would be, when Charles would come to meet me at Paddington; but that could not be, he had not had the grit to stand the intolerable winter climate of England, so he had gone away. And on that journey up from Plymouth how dull and wretched did I feel, the sight of the spring flowers in England, primroses and blue hyacinths growing along the railway line, almost brought tears to my eyes, as my thoughts wandered back to childhood, South Acre, and the Easton woods in the spring. All the others on the train were looking forward to meeting again with those they knew and loved; I had no one.

At my studio I found a whole pile of letters waiting for me, four from Charles, as dear as ever. In the last-dated one he spoke of having got fever, and of course this worried me, though I did not attach much importance to it, as he was full of returning soon to London. But the days slipped by and he did not come. In every letter he spoke of his return, though he still seemed depressed, and out of spirits. I could not send him money for his journey; my account was overdrawn to the tune of about £50. I wrote and told Charles this, also saying that he really must make a more economical arrangement for his mother, as I could not go on paying £52 a year just for a woman to go in and look after her, so she must make up her mind to live with her daughter, or some of her grandchildren. And alas! I added that if she could not do this, he would have to stay on there and look after her, 'much as I shall be grieved and disappointed not to see you'. But when my big annuity paid its first dividend I wrote telling him that as soon as my income had accumulated sufficiently we would go away together as in the old days, and I would give him £10 a month, which he could please himself about sending to his mother. Oh! how I was longing to leave England again, and go far away, alone with my dear Charles to some warm land over the seas; then we would be happy once more, and the wilderness would have no loneliness, and the desert land no desolation.

On June 14th he wrote that he was ill again with fever for two weeks. I was beginning to feel horribly anxious about him, and so I hesitated no longer, and wrote enclosing a cheque for £25 for his journey, second class, from Syria to London, only hoping he would not be too ill to travel. I felt a little comforted after I had done this, but his usual weekly letter did not arrive. What did I care that I was selling those wretched *Polydamas* at £3 a piece to Lord Rothschild? Money was nothing to me without Charles. If I had not taken on that hateful job for the old lunatic Longsdon none of this would have happened; when Charles became so unhappy in London I would have come back to him at once, or arranged for him to join me, wherever I might have been. And now a most pitiful little letter arrived just to say he was 'very danger ill' and knew that he was not going to get well.

Two days after his birthday when I got back in the afternoon to the studio there was a letter in my letter-box with a Syrian stamp on it; but it was not Charles' handwriting on the envelope. When I opened it, it was to read these words: 'Dear Madame, We are very sorry to inform you that Charles is dead after two weeks sickness. Before he died he asked us to continue correspondence with you & relating to us with the most consideration the time he passed with you in regard which you have done to him.' Then this letter proceeded to say that they would always like to hear from me, and to have my 'kind protection as a real mother', and it was signed 'All Charles Family'. Nothing could help or comfort me now.

TEN

——————— ❀❀ ———————

The Lonely Traveller

*Begging letters —— and Khalil's 'wife' —— grief and
blackmail —— sailing for Rio —— wandering life with
miserable persistence —— but still appetite for a pork
chop and a splendid view —— up the Para river ; a
butterfly-hunting clergyman in jeans —— up the
Amazon ; hotel with no beds but rings for hammocks ——
river monsters and poisoned arrows —— revelling in a
hot climate —— dying planter on a riverboat —— an
attentive Brazilian —— dawned on me that he would
soon be out of his pyjamas —— do antbites keep off
rheumatism?*

Miss Fountaine had been planning a collecting trip to South
America; she now told herself that if she concentrated on her
work and the arrangements for that journey she might set her
unhappiness at a distance. The effort failed; sleepless and ill, she
was unable to do either. 'Maybe the reader will think I showed a
great want of self-control and no courage,' she wrote, 'but I was
an old woman and Charles had been everything to me for 27
years; how could I face the lonely years before me?'

Her unhappiness was hardly diminished by a succession of
letters from Syria, purporting to be from Khalil's sister, his
mother, a friend. All had the same refrain; that she must keep in
touch with them, that Khalil had told them she would look after
them, that she should continue to send them money, that she
should sell his things left at the studio and send them that money
. . . Miss Fountaine noted that the letters from the women must
at least have been written for them, since neither could write or
speak English; she replied politely, she bundled all Khalil's
things up and despatched them to Syria; when they were gone
she felt the last link with his loving care had been broken. She
sent no money.

A further letter came from Damascus, from 'his best friend',
offering her sympathy for the sake of her 'holly friendship with
Mr Neimy . . . all your recent letters received by his people and

97

every thing is understood but even his wife does not know anything about his life because he was a very careful man . . . he
spoke to me of your friendship and his travilling with you and
the butterflies etc.' Miss Fountaine concluded that an attempt
was being made to blackmail her. She did not reply, but at last
got away from London and its sorrowful memories, and sailed
alone for Brazil.

Rio de Janeiro I had always heard of as the most beautifully situated city
in the world, and no description had ever come near the reality. Through a
beautiful bay of many islands one comes to a huge city, straggling all over
low foot-hills, with glorious forest-clad mountains towering above it on all
sides. The view from my hotel window looking over the city to the Sugar
Loaf Mountain, with the sea and distant mountains and islands, was magnificent.

My very first day's collecting in Brazil I returned from without a single
thing, not even an egg, a curious beginning to a country where I was going to
find more entomological work than I had ever known before. I secured the
services of a Swiss man, about forty years of age, Mr Meier, to accompany
me; quite a character in his way, with his queer, whimsical smile and quaint
way of surveying life, untiring in his efforts to help me, especially being
extremely clever about recognizing the different food-plants, which was most
useful. I am singularly deficient in botany, which it is so necessary for an
entomologist to study.

My one desire was to be so busy that I could never have time to think,
and indeed I was busy; at 5 am most mornings I would be up feeding caterpillars; then at 7.30 Meier would arrive at the hotel, and we would go out,
probably to find many more caterpillars and see yet other kinds of butterflies
laying eggs, till I really had almost more work than I could get through single-
handed, in the hours of daylight, especially when the bred specimens began
to emerge. But I rarely stayed at home, except on the days when I had a larva
to paint, and this was pretty often now, as I found the *Anaea* and the *Ageronia*,
of both of which I had several species, extremely beautiful, extremely interesting and extremely difficult to do; moreover, the different species were
most distinct from each other in the larval state, and I wondered if they were
all already known to science. But in spite of all my efforts, sometimes at the
end of a long day I would sit alone at my little table in the dining-room, and
maybe the band would be playing, and the music, the hum of voices, the
clatter of the dishes and the hurried footsteps of the waiters would all sound
distant and far away. Oh, why did I not die before him?

Then I got two more letters from Syria, both presumably from the woman who appears to have been Charles' wife; in the first of these she said, 'I beg you to pay attention to me in ordre to tell you about his strange biography. Since thirty years the deceased Charles has married me and I have born three girls and one boy to him and they are all alive. I am wearing in ordre to nourrish them . . .' The second, also (presumably) from Charles' wife, but in quite another handwriting and a different tone in the letter to any I had yet received, began, 'My dearest. Having not heard of you for a long time, we were expecting unpetiently to receive a word from you to amuse ourselves in our grate sorrow. According to my husband ordre, Charles when he understand his sickness was dengerous he told me that I might keep on writing to you from time to time and according to his advice I carry away his word. My husband Charles has four daughters and a little son, inclose please you find the picture of my smallest daughter which demand of God to brolong your life . . . My husband Charles did not mention to you that he was married and has a wife and children, the reasin is, because if you should know that he has wife and children you would not keep him so long in your part far away from his family, this is the way of every merciful person, and now are unpatiently waiting to heare from you and if possible to send me some money and to support my little ones and my husband's sister. Please forget us no and except our compliments Yours very truly Madam Naimy.'

I was quite aware these letters were more or less lies from beginning to end. Why should a woman at the end of thirty years of married life be left with a family of young children in their earliest infancy? She herself must be going on towards fifty years of age by now. Without doubt the whole thing was just a scheme to try and get money out of me, and as to any statement my poor Charles may have made on his death-bed, I did not believe there was a word of truth in any of it. I did not answer these letters.

During her stay in Rio, Miss Fountaine was attacked by two large dogs and suffered an eight-foot fall, injuring her back; undaunted she set off for a month's collecting in the mountains, complaining only that 'Brazil was a hateful country for an old woman travelling alone. The Brazilian has no use for an aged ♀' (she uses the zoological symbol for female) 'and for the young he is even more repulsive with his disgusting gallantries.' This unkind view may have followed from her lack of the local language; with the educated Brazilian she would get by on English, French, German, Italian or Spanish; her disgruntlement with the Brazilians fades a little as, with time, her Portuguese presumably improves. In her travels she was never daunted.

I expected to find a motor-car at the station to meet me, but a tall, dark peasant man appeared with three rather miserable looking horses. Could I ride? he seemed to be enquiring. 'Yes,' I answered, 'but not in these clothes', the skirt I was wearing being exceptionally narrow. Every time I announced my intention to walk from the station to the pension the alleged distance increased; till at last a woman in a store near by got wind of my dilemma and, motioning me to step inside, took me into a back room and fixed me up with an extremely voluminous black petticoat, in which I certainly could ride; and once on horseback I felt as right as rain.

After a month in the mountains she returned to Rio, still delighted with the beauty of the city, staying there six months instead of her planned six weeks. Eventually she moved on to Recife in Pernambuco, noting that, lacking the zest of earlier years, she had 'only a sort of miserable persistence which compels me to continue the same old wandering life from force of habit'. But, she confesses, her enjoyment of nature, of tropical forests, birds and insects, was greater than ever. She complains, on 15 April 1929, that she must write or 'by tomorrow my wretched memory will recall little or nothing of the events of today . . . but I may as well go on with my story now to the end'. She was still able to record, however, that she enjoyed a day's collecting, including a magnificent blue *Ageronia* female, and returned to her hotel in Pernambuco with a country appetite, to enjoy a cup of cold consommé, a delicious pork chop, and a splendid view.

Her next hunting ground was to be at Belem, on the Para river, the home then of the Reverend Miles Moss, one of the world's great butterfly experts. She had long wanted to meet him, though someone who knew him had 'to an extent prepared me; shall I say, disillusioned me'.

I had fondly pictured him as a tall, rather ascetic looking parson, deeply mystical and not by any means uninteresting, so when a thick-set, rather short, middle-aged person, not looking like a parson at all, rushed in on me one evening shortly after my arrival I was not so very much overcome by his appearance and almost boisterous manner. Of course, we at once began to talk 'shop', and a kindred spirit . . . well, *is* a kindred spirit, no matter what its external attributes may present to the (perhaps) too critical eye. And before he left an expedition was arranged for the next day to Utinga. Many were the days I passed in this lovely place, where the grassy earth was sodden with rain and a good deal of the forest was a swamp with dark pools where water-lilies flourished, and where the dense, heavy jungle lay sleeping in the calm, dark water, in that intense and absolute stillness so truly tropical and to me so truly enthralling.

However, to return to that first day, I must admit to a slight shock when I saw from my window in the hotel a stout middle-aged person clad in a complete suit of old blue dungarees, get off a tram car. I thought I must have been mistaken in recognising him as my friend of the night before, but he made straight for the hotel, and came up in the lift, no one seeming at all taken aback with his dilapidated appearance. It passed through my mind that by comparison my own country outfit would seem quite respectable. And I soon got accustomed to the old blue dungarees, and everything else connected with Mr Moss; indeed without his aid I should certainly have found this district most difficult to work in, apparently so little deserving of its world-famed entomological reputation. Nor was he only a first-rate entomologist. His drawings of the caterpillars were the best I have ever seen, and put my own efforts in the shade; and besides this he was a first-rate musician, playing the organ himself in the little English church standing in its own grounds where he had planted many beautiful and interesting plants and shrubs, some specially selected as being the food plants of certain cater-pillars. It soon became an understood thing that every Sunday after the morning service Mr Moss would come and lunch with me at the Grand Hotel. My room was soon full of caterpillars, till every cage and every bottle was crowded, while most days would bring me some new treasure from the jungle, very often found and given to me by Mr Moss. It was very hot but I revelled in the luscious climate and loveliness of this tropical region.

I left Belem by boat up the Amazon to Obidos, with a Mr Harding, a white-haired Irishman of sixty, as my guide. The hotel at Obidos, little better than a dirty, very uncomfortable inn, had no beds at all, only large rings in the walls on which to hang hammocks. Luckily Mr Harding had come provided for this, and I had my camp-bed. The sanitary arrangements were quite impossible to dwell on. We took another boat on up the Madeira River to Porto Velho. The boat stopped to take on a cargo of cattle; they were fed on a kind of coarse grass for which the boat would from time to time stop near the bank. Men would go on shore and cut the grass for them so that what with this, and stopping for the wood the boats burn, it was very evident that the Amazonian river boats would never be in a hurry.

The Madeira is an immense tributary of the mighty Amazon, navigable for about eight hundred to a thousand miles, and here indeed the rich Brazilian lepidoptera displays itself beyond the dreams of the most sanguine entomologist. In some places the wet muddy banks were lined with butter-flies, which from mid-stream were plainly visible as large, greenish yellow patches, where thousands of these gay, beautiful butterflies had congregated to imbibe the moisture. Others often flew round our boat, but were most difficult to catch; even more wonderful were the moths at night, attracted by the electric lights. If I got up early enough the next morning before the man had been round to sweep them all remorselessly away, I could just pick off all I wanted.

In the immediate vicinity of Porto Velho the matto had all been cut down. There was, however, a fine forest within an easy ride of the little town, and it was here that Mr Harding and I used to go with two boy attendants. As soon as we arrived in the outskirts of the forest I would dismount, and, retiring to some secluded spot, would change my riding knickers for an old skirt, always preferring female attire as being so much cooler, especially as I had with my usual stupidity invested in a substantial pair of the before-mentioned article of clothing at Harrods, made specially for women to ski in during the winter in Switzerland. How I could ever have supposed that they would be suitable for riding in the tropics is to me now quite incredible.

The weather was fine and the multitude of butterflies I saw and sometimes caught never to be forgotten; unfortunately I had not yet learnt how to suspend bananas soaked in sugar-cane juice as baits, so that the only special attraction were the fresh droppings from our horses, and on this mess I sometimes caught *Prepona* and *Anaea* with my fingers. I got my boxes crammed with lovely specimens. Mr Harding used to say nothing would he like better than to see every particle of these magnificent forests in Brazil cut down and burnt. I told him that, could this happen, the rainfall would be considerably diminished, that the waters of the Amazon would have to be used for irrigation, as the Nile waters the deserts of Egypt. Yet he loved all the living creatures of the jungle, and often I would find him surrounded with butterflies which were settling on his hands and any other part of his person which happened to appeal to them. These 'pets' of Mr Harding's were sacred from the ravages of my net – though if anything really good had ever been amongst them, I can't answer for what might have happened.

They travelled on, further into the interior, crossing briefly into Bolivia; Miss Fountaine also did some collecting on the river bank despite her fear of the hidden muddy depths and the stories of man-eating piranha fish and of mysterious river monsters. Once when she was on the slippery, hazardous bank, she saw 'some huge monster flopping about close to the brink where I was about to descend, but it disappeared into the depths before I could get a good look at it to determine whether it was fish or reptile'. On another occasion she pressed on alone deeper and deeper into the forest, looking for a pool of drinkable water and a rich source of butterflies which had been described to her.

It was wild enough to be to her liking, she writes, though even she began to feel 'a bit creepy'. Her escort followed her reluctantly and galloped back madly ahead of her as she returned. Only afterwards did she learn that the forest was the haunt of an Indian tribe who had recently murdered several white people. 'The risk of being shot in the back by a poisoned arrow would not have added to my enjoyment of that wonderful forest, had I been aware of it,' is her only comment. There were other macabre moments of which Miss Fountaine gives us brief, almost cinematic glimpses.

Before we left Porto Velho a planter in the last stages of malaria had been carried on to the boat on a stretcher; his wife and small child were with him. He was being taken to the hospital in Manaos, but he never reached there. One morning Mr Harding came to tell me that the poor thing was dead, and all that day we knew they were busy constructing a coffin, first having had to stop somewhere to get suitable wood, so that it was quite dark before we reached the place where the interment was to take place. First there was the usual arrangements of boards placed from the deck on to the muddy bank of the river, and then a wide stretch of mud had to be traversed to reach the foot of a long wooden staircase, of some fifty feet or more, to the top of the cliff. It was a dark, wet night with no moon, and this wretched little torch-lighted cortege slowly wending its way through the rain and the darkness was one of the most gruesome spectacles I ever beheld.

Nor were riverboat funerals the only disturbing scenes. Alone in one riverside hotel she fell into conversation with a Brazilian, 'apparently of a most friendly disposition, and though I had certainly remarked a somewhat unnecessary tenderness in his manner, and seen a look in his eyes which, though familiar enough in the days of my youth I had not seen on the face of any man for years; still it was only after he had gone away to his room from time to time, returning again and again to the sitting-room where I was waiting, that I presently became aware that he was now in his pyjamas, and the next thing that forced itself upon my obtuse, unsuspecting brain, was that very soon he would be *out* of his pyjamas. Seeing, therefore, that it was getting late, before he had time to see what I was going to do, I wished him a good-night and went to my room, taking care to lock the door after me. It reminded me of the days of long ago, but in a way I did not wish to be reminded, least of all with a beastly Brazilian.'

All through this little scene a storm had been raging; wind

rising to hurricane force and rain in a deluge. Fortunately it had calmed considerably before she had to set off in a canoe to meet her steamer in mid-stream – a greater source of anxiety to Miss Fountaine than amorous Brazilians. This steamer was the first stage in her return to England. Regretfully she wrote, 'It is all over now, the long, steaming hot days tramping through wet forests with water sometimes up to the knees, but always with a warm luscious air, even in the blinding rain, day after day. I would return soaked, but warm rain never does harm me, though possibly the incessant bites of very formidable ants may inject so much formic acid into my system that any signs of rheumatism are at once checked.'

She was fit enough when the boat reached Madeira to be amused at 'the way everybody seemed to think I was a sort of antiquated old party, who needed looking after. Doubts were raised as to whether I should care to come down from the heights above Funchal, in one of those little sliding open cars which descend at a head-long rate, with sharp twists and turns; I thoroughly enjoyed the whole thing.'

After a brief stay in London she was off to South America again, this time to British Guiana. She shocked her (female) guide by drinking iced beer with a Dutchman and a Portuguese at a hotel somewhere up the Demerara River ('the topic was the origin of the Albino'); she penetrated to the diamond-mining town of Bartica, which had so wild a reputation that another, male, guide was afraid to go with her. Instead, she hired an Indian, who helped hack a way through the forest with his cutlass. 'He was always whittling with his knife at something or other, and when I asked him if he was making blow-pipes the idea seemed to appeal to his sense of humour, and he would laugh a sort of laugh, and tell me he was merely preparing these sticks to make fans; nothing more deadly than that. Sometimes I would observe him busy collecting the leaves of a certain shrub, and these he told me would be made into a kind of poison to catch fish, though I could never make out that he had succeeded by this means in catching any, as it seemed a long process was first necessary in its preparation.'

She stayed with an English rancher's young wife at a remote cattle station, going for long rides with her until 'I had to re-member I had not come to British Guiana to spend all my time galloping over the open savannahs, however delightful that might be'. She returned to Bartica, seeing there some of the effects of the great slump with which the thirties began: 'The depression in British Guiana was very great, owing, I believe, principally to the slump in diamonds,' she writes. 'Whole boat-loads of the wretched diamond miners would arrive in Bartica, and I shall never forget one Sunday afternoon watching from the windows of the hotel, when a number of these poor things were arriving, the pouring rain drenching them and their miserable, small bundles. The bedraggled wretches had come down the Essequibo river in an open boat and had had little or no food for several days. Unsheltered and unfed, was it to be wondered at that one man had died on the way and two more succumbed afterwards in the hospital at Bartica? And all this was going on in one of *our* colonies!'

There spoke a voice Queen Victoria would have recognized.

———————— ❧❦❧ ————————

ELEVEN

Organising the Family
1929

*An escaped convict on the Orinoco —— taking to the air,
with caterpillars —— joys and sorrows of the American
Fountaines —— putting the family to rights —— leaving
in style for New York —— the Great Crash and no
more music and dancing —— the boat to Cuba ——
45-mile horse ride at 69 —— an earthquake foretold by
a dog —— England, servants and equality —— to
Kenya, remembering South Acre and sparrow pie ——
buying a £50 car in Uganda —— the young Scots
botanist —— If I were 50 years younger . . .*

If nothing else were to mark out Miss Fountaine from other
ladies born in 1862, it was the willingness, indeed the eagerness,
with which she responded to the possibilities of flying . . . travelling, as she put it – perhaps because she had first given serious
attention to the idea when she was in France – by 'avion'. Then,
problems with luggage and her fear of Khalil's disapproval had
stopped her. Now, when her travels brought her to Ciudad
Bolivar in Venezuela with no easy way out again, she hesitated
no longer. Her guide in the area, a M. Réné, and her caterpillars
went too.

After a good deal of unsuccessful bargaining with the chauffeurs of
automobiles, I found it would be much more desirable in every way for me
to go to Upata by avion. So, having despatched our larger luggage by camion
(lorry), Monsieur Réné and I, together with two or three other passengers,
soared away up into the clean sky, leaving the squalor of Venezuela far, far
below; but it was a very small avion, and not covered in, so that the noise
and the wind were terrific. I had climbed in, not without some difficulty, to
sit with one passenger just behind the pilot, and under the wings; the
higher we went up the stronger the wind became. The trees and the houses
and the people were but a vision of something, to which we up there in the
pure air and the brilliant sunshine were a thing apart. Once, far below, the
broken fragments of a blue river were to be seen, and not long after that my

fellow-passenger told me that Upata was in sight, a statement I was not altogether loth to receive, for though I had thoroughly enjoyed it, and hoped though it was my first that it would not be my last flight, I was beginning to feel the restrictions of the very limited space into which I had been literally wedged.

Automobiles were here to meet us, and it seemed a considerable portion of the population of Upata had also turned up to witness our arrival, for the avions only come here if they have at least two passengers for Upata, so on this occasion it had descended entirely and for no other reason than to deposit myself and Monsieur Réné, and our smaller grips. The caterpillars and pupae I still had from British Guiana had gone up with me into those high altitudes; one *Laretes* pupa was on the verge of coming out and very nearly did so up there in the air. I wished it had.

> When she came to move on – Upata had not been rewarding to a collector – there was no aeroplane to take her the thirty or so miles to the nearest port on the Orinoco and no cars available; she settled for a very old truck laden with bullock hides. Her luggage was perched on top, she rode in the cab, while the unfortunate M. Réné and another man stood one on each running board. 'Why that damned thing never turned turtle was little short of a miracle,' she wrote. 'Once it jibbed hopelessly on a hill and we found ourselves retreating backwards till one of the wheels got wedged in a ditch and this at least brought our wretched vehicles to a standstill.'
>
> Finally arrived at the little town of Tucopita in the Orinoco delta, she talked with convicts escaped from French Guiana ('a most harmless looking little Frenchman, I could not think that his crime had been a very ghastly one'), and watched the Good Friday procession ('a gruesome effigy of a dead Christ . . . a rather distressing band played Chopin's Funeral March over and over again'). A few days later she was in Trinidad to celebrate her diary day for 1931. She complains it began with the discomfort of menopausal hot flushes . . . 'This has now been going on with me for over twelve years, without so much as one day or one night's respite; and next year I shall have reached the allotted span of human life. Surely I am beating all records! It was eight o'clock before I got down to breakfast, a good, substantial meal it was too, as I am still enjoying the hunger subsequent to weeks of semi-starvation in Venezuela . . .'

On the slopes of Trinidad's beautiful mountains is a monastery called St Benedict. The order is a strict one, no female is allowed to enter its precincts. What I think of this arrangement, by which a number of men are all herded together, need not be expressed. Father Maurus, one of the priests in this monastery, an entomologist, was one of the most pure, simple-minded men I have ever met, a genuine lover of nature, a true scientist, with the simplicity of a child. A most congenial companion, he was allowed to go out with me pretty frequently, perhaps because I was a somewhat antiquated specimen of my sex; but on no account whatever must I set foot in the monastery grounds, much to the regret of Father Maurus, as it would have been a real pleasure to him to have shown me all the treasures in his laboratory. He was still young, not much over thirty, and almost painfully thin. His activity on the hillsides and in the ravines must have been somewhat hindered by a long, light coloured garment resembling a cassock that he always wore, but I suppose he was accustomed to it. And we certainly went into some wild places, when maybe I would wade up to my knees in a mountain stream, threading my way between the huge rocks and boulders in the wake of the boy George who generally came along with us, while the lean, spare figure in the light coloured cassock would be climbing over the rocks above.

There was peace up here in the unpretentious little guest house just outside the monastery grounds; but when I would wake in the early morning to hear the chanting of the priests, as some rite would compel them to make a procession, I would wonder how Father Maurus, with all his love of nature and the true God whom one approaches so nearly through the marvellous beauty of creation, could bring himself to believe in such a grotesque exhibition. I enjoyed my rambles with him over the mountains, but I did not get much of interest owing to the dryness of the season. I am so discouraged by the hopeless failure of my entomological work for many months past that I sometimes feel my day is over and I will never be able to get 'our' collection up to 20,000 specimens.

> For some time past, Miss Fountaine had been getting 'most miserable' letters from her brother Arthur: he and his wife Mollie had now parted and Miss Fountaine had promised not to re-cross the Atlantic without visiting him. She had intended to put off her visit until the following summer, but 'I now decided to go to him at once, feeling that if I did not do so, it would be too late, for his description of his health was lamentable'.
>
> Despite the obstructions of American immigration procedures, and the delays of Customs men ('None of the coarse insolence of the Venezuelan here, but a most inquisitive turn of

mind . . .'), she finally got into New York, to Washington (the White House was magnificent, she noted, but did not approach the King of Siam's state apartments for beauty) and finally to Covington in Virginia, and her brother.

From the sound of his voice on the 'phone, I was perhaps a little prepared for the pitiful sight he presented. A very old man (he might have been over eighty, instead of not much over sixty), with perfectly snow-white hair and very little of it; his voice was shaky and trembling, and he mumbled terribly, though this defect must have been considerably aggravated by the fact that he was entirely destitute of teeth. The poor thing seemed quite overcome with emotion, but genuinely pleased to see me, and from that moment I was glad I had given up everything to come to him. He was suffering from hardening of the arteries, not an uncommon complaint amongst old people; and though no doubt in the case of poor Arthur he was paying the heavy penalty due to the excesses of his youth, his sufferings at times were heart-rending. He was astonished at my strength and vigour, which I must own, thank God, is rather remarkable for my age, and perhaps it was scarcely kind to tell him that it was owing to the healthy life I had led.

One can see Miss Fountaine irrupting into the peaceful community of Covington and into the lives of her kinsfolk and their neighbours. Time and trouble may have tamed her a little, but she was still the tall woman of commanding presence and Victorian confidence who had terrified British consuls and Balkan Customs Officers, muleteers, Circassian bandits and the attendants aboard innumerable ships and trains. She surveyed the scene and was not pleased. Her nephew Lee was living out of the town with a farmer; she expected him to visit her and when he did not she went in search of him, finding him, late one evening, boarded out, overworked and browbeaten in a tumbledown farmhouse. He was as beautiful a boy as he had been a small child; she longed to take him, shy and sad, into her arms now as she had then. And the farmer would not allow him to go! Lee, her nephew, one of the few remaining male Fountaines! 'The boy would be as beautiful as a Greek god had he been brought up in refined surroundings instead of among a mob of Virginian farmers,' Miss Fountaine snorted.

Then there was Mollie, whom Arthur had been so fortunate to marry, now estranged; Miss Fountaine talked to her, rode with her and with the other nephew Melville, and decided Arthur's doubts about his wife were absurd.

'I saw that I had indeed not come to this country for nothing,'

she remarks grimly. She succeeded in very short order in re-
uniting Melville with his father, and in detaching Lee from the
brutal farmer.

At this point the farmer's wife and another woman were foolish
enough to visit Arthur with a claim for damage to a cow allegedly
hurt by Lee; 25 dollars, and 100 if the unfortunate beast should
die. As one who pursues a domestic cat and finds he has captured
a tiger, these ladies were faced not by the sickly and gullible
Arthur but by his formidable sister. One woman – 'a horrible
creature with a face enough to turn all the milk in her dairy sour,
was so overbearing and insolent that any idea I might have had
of helping out was immediately quenched'. The other was soon
'reduced to tears, leaking copiously in the public lounge of the
hotel. Englishwomen have given up weeping now, they swear
instead, and I surely did swear.' And the woman and their cow
disappear from the diaries never to reappear, nor is there any
word of money for them.

Both of the boys loved motoring, so their aunt hired a car for a
week in which the boys drove her about the countryside; later
she bought them an old Ford ('a roadster, they said it was'), from
which aunt and nephews alike had a lot of pleasure.

By the time her four-months visa was about to run out, the
American Fountaines had been thoroughly shaken up, and put
back in a way more agreeable to their visitor. Even Arthur was
better, 'almost entirely because I had persuaded him to give up
drinking coffee from morning to night'. One feels it would have
taken a braver man than Arthur to continue ailing in these
circumstances.

And when she left, she left in style; American go-getting style
– it was the English spinster who chose to cut eight hours from
her journey by flying to New York, while her American brother
thought it nothing short of suicidal. She ignored him, of course.

They all, including Mollie, came to see me start off from White Sulphur
Springs on my flight to New York. As I soared away above the wooded hills
surrounding the Air Port, and I looked back at the little party standing there,
Arthur and Mollie and the two boys, I felt a great love for them all.

It was quite a small plane, and I was the only passenger, but no early
Victorian lady in her sedan chair could have travelled in greater comfort, or
safety. I became more convinced than ever that this would be practically the
only mode of travelling in the future, so that Lee and Melville as old men
will be looking down from the air on overgrown, disused motor roads; and
people will be saying to each other: 'Just think how dull it must have been

in those old days, crawling about over the surface of the earth!' I was not looking down on disused motor roads, in this year of 1931, anything but it, for cars were skimming along their paved surfaces, no doubt going at a great rate, but compared with the soaring flight of the plane some thousand feet or more in the cloudless sky above them, they sure did seem to be merely crawling along.

At Washington I managed to send a wire to Arthur, and snatch a hurried drink, before the big plane for New York was ready to start. Washington from the air was a sight to be remembered, and by and by we passed over Baltimore and Philadelphia, and I loved to find myself up there in the blue sky away from all the frivolities and miseries of those crowded cities. Of course, we put down at their air ports, and a very bumpy process it generally was, and each time we descended to earth I longed for the moment when the plane would soar away again into the far blue sky. I had left the White Sulphur at 2 pm, and at 6 pm we arrived at the air port of New York, then only forty minutes by motor-bus to the Pennsylvanian Station.

> She was in New York on the Monday when Britain went off the gold standard; the devaluation of the pound which followed was a blow to Miss Fountaine's financial position. She cut out lunches altogether, and stopped taking taxis, no easy sacrifice in a city she didn't know with distances too great for easy walking. She took some comfort in the thought that the United States, 'though choked with gold', was also in a bad way. Banks were failing all over the land; the American woman with whom she had been going to visit Cuba lost nearly everything in such a crash and had to call off the trip. Miss Fountaine dreaded travelling alone, but set off just the same.

It was a big boat, but a mere handful of passengers (22 in all) made it very depressing. That spacious deck and all those beautiful reception rooms were empty and silent; the laughter, the whispers of flirtatious couples, the sound of music and of dancing, was no longer there. In Havana it was the same miserable story of empty hotels, and fine shops with everything in them but customers, men loitering at street corners with apparently nothing to do but discuss politics.

I decided to go to Santiago by avion, sending my luggage by rail. This was the longest trip by air I had yet made – five-and-a-half hours, and as my guide book said, delightful. I had looked forward to a bird's eye view of all those wonderful Cuban forests I remembered here 21 years ago, but alas, all I saw was a sort of chess-board effect of square fields, cultivated with sugar cane. We alighted and stopped a few minutes at several places, including

Camagüey, but there too, the forest was all gone as though it had never been. The pilot, a young American, told me of a very nice boarding house in Santiago, and after a rough flight – I began to feel faint, so I took a sip of brandy I of course had with me, and soon felt all right again – he took me there himself.

I spent most of the winter inexpensively in Quantanamo, for the exchange was going more and more against sterling, in spite of the fact that the October General Election in England had resulted in a big Conservative majority. There was nothing for it as far as I was concerned, but to spend $3.20 where I would have been spending $4.80. Perhaps this enforced economy was good for me; maybe I had become too purse-proud and extravagant, and now having gone to the other extreme I realise that neither conditions are so great as in my rather unbalanced imagination I had depicted them to be; and anyway the restrictions as regards food are not doing me any harm, quite the reverse.

> She enjoyed her stay; the town was the home of a Dr Ramsden, a fellow-collector, 'one of the very nicest men I have met, even among entomologists', and of a Spanish girl entomologist with whom she went collecting and riding.

On one trip two peasant men, who we had often watched ploughing industriously with a yoke of oxen, called us into their field to ask me if I could tell them what it was that was devouring their young cabbages. When I saw the devastated plants it seemed to me to be more the work of slugs, and the small larva of some insignificant moth than of butterflies. They were such simple, well-mannered men, and so confident that I could help them; so I told them to water their cabbages every evening with a solution of common salt and water. Some time later I saw as fine a crop of cabbages there as one could wish, but whether this was from their having followed my advice or not I am unable to say.

Another day we rode to a place $22\frac{1}{2}$ miles from Quantanamo, where we collected, or rather tried to collect, for there was very little to be got; then we rode back, galloping most of the way for our horses were both as fresh and keen as when they started. This was 45 miles in a day, and I am only four months from my seventieth birthday! I feel more safe and secure in the saddle than ever; indeed, when I am on a good horse I feel young again.

One evening not long after this, when we were playing bridge as usual at Miss Ashurst's, I could not help noticing the strange manner of their dog Jock. He seemed worried about something, and was never still a moment. That night I could not sleep; I felt the strangest foreboding, and I was still wide awake when soon after midnight the whole building shook as with an earthquake. The hotel was largely composed of wood, and I felt little or no fear, not even when the lights went out. I could hear the young American woman in the next room being consoled in the arms of her Cuban husband, evidently in mortal terror. But it was over, and I was soon so sound asleep that another, though not quite so severe shock at about 5 am failed to wake me. The earthquake had wrecked Santiago. Everybody now seemed at high tension, always in anticipation of another shock, but Jock and I knew the worst was over. It is strange how animals always seem to know when an earthquake is coming. I afterwards heard that the horses in the corral had also been greatly disturbed that evening, and when the shock actually came, they all went down on their knees till it was over.

> She was back in England for 15 April 1932. 'Cold! Cold! Cold! That was my first impression when I woke up . . . it takes me a long time to dress now, with so many clothes to put on. Browning must have been suffering from loss of memory when he wrote "Oh! to be in England, now that April's there". I feel a longing to be away in some land of sunshine and hot rain.' Meanwhile, she disliked carrying coal upstairs for her studio stove, and ashes down:

I sometimes wonder if I should dislike all this as much, were it not for the detestable habit among my grandmothers and great-grandmothers of considering themselves too delicate and too dignified even to replenish their drawing-room fires; whenever fresh coal was needed, they must perforce ring for a footman to put it on for them. Poor helpless creatures, with what pity they do inspire me – and sometimes I even pity myself for having sprung from such a breed!

I am now occupied with clearing the drawers of one of my old cabinets, which I am sending away in exchange for another new mahogany one. Today I had got as far as the drawer containing the *Agrias*, those very expensive but exquisitely beautiful things, and these were immediately followed by a few brilliant *Preponas* from the forests of Matto Grosso, which struck me as being more lovely than ever, and give me such a longing to be back in Brazil. The contents of the next drawer took me back to India and Java, for there were the *Kallimas*, then came the *Charaxes*, full of hot, African sunshine . . . My next trip I think will be to Madagascar and I thought so more than ever after spending some time studying Seitz's *African Rhopalocera*, where I find this big island mentioned many times, and generally with reference to most interesting and strikingly beautiful species. I only hope there are horses there, but I doubt it. Soon after three o'clock I began to prepare to go out, having been told that *Richard Arrowsmith* is a 'talkie' I should particularly like. The scenes depicting the plague-stricken island in the West Indies were wonderful, and the Negroes did their parts true to the life and most realistically, seeming to enter into the gruesomeness of the whole thing as though they rather enjoyed it, which I dare say they did. The tropical vegetation and (to me) familiar plants all steaming under a deluge of drenching hot rain was also most realistic . . .

No need to be content for long with memories and movies; soon Miss Fountaine was booking her steamer tickets, packing her trunks, for Madagascar. But the hunter was becoming more soft-hearted, and the prejudiced traveller less insular, as the years passed. In Madagascar she was able to admit, 'The more I saw of the people the more I found to like in them; they have something of French politeness . . . neither are they by any means lacking in intelligence.' As for the butterflies, when she caught a somewhat damaged female *Hypolimnas*, very tame and good, but 'could not seem to strike her fancy in food plants', she let her go; and when the eggs of a *Cyrestis Elegans* from a mulberry tree hatched, pupated and went through the transformation to butterflies, Miss Fountaine felt 'dreadfully sorry to have to kill such pretty, delicate little creatures as soon as they emerge, so weak and helpless'. She even felt affection for the pigeons which flew freely in the big dining-room of her hotel, strutting about to pick up the guests' crumbs: 'strangely enough they never seem to foul the floor'.

She wrote, as she sat at the end of her next 'day', 15 April 1933, that she was no longer an actor in the drama of life, but a spectator. She sat for an hour while smoking cigarettes 'very quietly and pleasantly', before retreating to her room to slip into bed under her mosquito net; perhaps, she wrote, to dream. 'Sometimes I dream that I am young again, but never feel any regret when waking comes and I know I am a lonely old woman.' For she found it consoling to think that she would not need to wander alone, without Khalil, for very much longer.

Meanwhile, as the next year's diary began, she was as strong and well as she had ever been, and walked twelve miles on the day after her 71st birthday. A little later, running one cold day to keep warm, she covered three miles in thirty-one minutes. In such a mood she sailed for Kenya and Uganda. She was still unhappy about her quarry.

I never before felt more sorry for any butterfly I have ever bred than for a poor little *Dardanus* female. She stretched out her long proboscis, and seemed to be feeling about to find something to suck – and I? I gave her petrol, till she died. It recalled to my mind an incident in my childhood long, long ago, when Papa had fed a place in the meadow at South Acre (I could show the very spot now!) for sparrows to come with confidence and assurance, till the mid-summer holidays, then to be caught by John in a net and killed; sparrows were somewhat injurious to Papa's crops, and John was apt to become very troublesome if not constantly amused during his holidays. That day we had made a big haul, and I had just taken a wretched little hen sparrow from the net, and was about to kill her, when she looked up at me with such a piteous appeal in her bright, anxious eyes, just as though she would say, 'I know you are going to kill me! Oh please have pity!' that for a moment I was moved to pity, but I dared not let her go for fear of John. The next moment my small fingers were pressing her little wind-pipe till she was quite dead, and the little corpse was put aside amongst the rest for the sparrow pie, of which there was little enough to it, except only the piece of mutton that was put in with them. And still the look of that helpless little hen sparrow has come back to me through life from time to time but never quite so vividly as on that day; and now that *Dardanus* will always be there in my collection to remind me of the pangs of remorse I felt before I took her life.

In Uganda, unable to get a horse, she spent £50 on a second-hand Ford, and hired a native chauffeur. All the more distant forests were easily reached now, and produced interesting and beautiful butterflies. The old car took her to Lake Albert where she spent 'the happiest time for many years', not least because of her friendly reception by a young Scots botanist who helped her penetrate the jungle and find the right food plants for her caterpillars. 'Had I been some forty or fifty years younger I should certainly have fallen in love with him,' she admitted, adding that it must have been a comical sight, the very tall young man striding ahead, and an old woman with her butterfly net pushing and struggling through the thickets after him. But the collecting was wonderful, and young Mr Eggeling couldn't have been more attentive if she had been nineteen; he even climbed trees to collect plants for her.

When *Love Among the Butterflies* was published in 1980, telling of Miss Fountaine's earlier travels, the Norwich Castle Museum, which holds the diary manuscripts, received a letter from Dr W. J. Eggeling of Trochry by Dunkeld, in Perthshire.

'It was in 1934 that I met Miss Fountaine . . . over forty years ago! – and I can still remember her with the greatest affection, as if it was yesterday,' he wrote. 'I was then in my mid-twenties, just starting my second tour as an assistant conservator of forests in the Uganda branch of the Colonial Forestry Service. I do not remember how it started, but soon the whole verandah was covered with breeding cages, and Miss Fountaine busy from dawn to dusk, collecting fresh food for the caterpillars, collecting, mounting or painting. She may not have been a great scientific lepidopterist – whatever that may mean – but I doubt if anyone to this day has unravelled so many butterfly life histories as she did, or collected such wonderful material, perfect because the butterflies were so often hatched out in the cages and killed before they suffered any wear at all.

'I don't think it will be too much to say that I loved Miss Fountaine, certainly I had the greatest affection for her, and the greatest admiration. She was, so it seemed, devoted only to her butterflies, and to travel, to see new places. She never spoke of the past and of all she had done, and I never suspected she was other than a Victorian lady straying very late into the next century. I suppose I was young and naïve . . .'

The Castle Museum was able to send Dr Eggeling photo-copies of the pages of the diary, unpublished until now, covering her trips to Uganda. In reply he wrote: 'It was very good for my ego, as well as touching, to read that Miss Fountaine would certainly have fallen in love with me had she been 40 or 50 years younger – and no doubt it would have been reciprocal! However, having read the book I realise she fell in love quite often!'

TWELVE

————— ✪ —————

Faster and Faster
1935–39

*A drink straddling the equator —— home and off again
—— a snake in the car and flamingos on a sunset lake
—— reflections on girlhood chances —— the USA and
her nephews; playing pool in the saloon —— young Lee
in a shack; Miss F. buys him a farm —— coarse jokes
about Mrs Simpson and the King —— alas, no air
service to Saigon —— rickshaw through a storm and a
tour of Cambodia —— Angkor Vat's ruins and Hanoi's
flower-filled boulevards; my best season —— Christmas
with a family in Singapore; paperchains and paper hats,
butterflies and children —— writing up her diary
'perhaps for the last time' —— an early start —— and
the rest of the book is blank*

Before her year was up she had been marooned in her broken-down car in the forest (looking in vain hope for the dangerous wild animals against which people warned her) and had drunk a glass of sherry in a bar straddling the equator, taking part of her drink in the northern hemisphere and the rest, further down the bar, in the southern.

She returned to England, regretting the cold even before her ship was through the Mediterranean. England was as cold as ever – what did these people mean when they spoke of the summer 'heat' and were in a state of exhaustion? It was pleasant, though, to meet her old friend Skye, whose house in Park Lane had been the scene of such glittering entertainments; Skye was nearly 90 now, but still with a clear intellect and an insatiable love of croquet. And it was pleasant to drive out to Goodwood with young Captain Riley from the Natural History Museum and watch his children chase butterflies where she remembered, so many years before, galloping with another girl on hired horses from Chichester. So many years before . . .

Early in September she was off to Africa again. The pace of the diaries seems to become faster and faster; with no personal ties any longer she is recording only places, movement. The sudden death from blackwater fever of a man she was due to meet drove her to speculate on the way human life is governed by chance – and in speculating she fills in some of the blanks in her record of earlier years. 'For instance,' she wrote,

had it never happened that Hennie Guise had obtained that invitation for Rachel and me to stay at Colesborne, so that by making the acquaintance of Henry Elwes I should find myself a few years later waiting for him and Mrs Nicholl to join me in Syria (a country which otherwise it would never have occurred to me to visit at all), I should never have met Charles, and the whole course of my life would have been absolutely different. I suppose I should have married Mr Rowland-Brown; under those circumstances he might perhaps never have died from over-work during the war; and all this from the simple fact of Hennie Guise getting us that invitation.

She bought another old car to tour the country again; she watched native kings crowned, saw – as appeared to be the stand-ard experience of motoring Africa – a snake take refuge on a car engine, and experienced an earthquake ('old Miss Barnes having been convinced that the shock was caused by elephants trying to push the house down, this surmise had promptly thrown the poor lady into hysterics').

She escaped unharmed from a puff-adder (she had been too busy searching for caterpillars to notice it a few inches from her ankle), and lay in bed listening to a lion growling outside her room; she met a couple walking the length of Africa provided only with autograph books – they succeeded in getting overnight accommodation, meals, and on occasion enough to get drunk on, by cajoling District Officers to add their signatures. She met her charming Scot again, and a titled lady touring Uganda with a chauffeur, who – Miss Fountaine observed from the next apart-ment – shared his employer's room until the early hours. And she saw – and it thrilled her to her patriotic heart – the celebrations of George V's Silver Jubilee, there in the heart of Africa.

The longer she stayed in the long-gone, peaceful, prosperous Uganda the more she liked it. But when after many months her mail from England caught up with her, she learned that her brother Arthur had died. As one of her nephews' trustees (her

sister Evelyn was the other) she would need to return and attend
to business. But not yet; not while there were new sights to see –
the Murchison Falls, the beginning of the Nile, Lake Victoria,
the peaceful little townships of Kampala, Jinja and Entebbe. Not
while there were young entomologists now eager to seek her out
and listen to her voice of experience; not while there were the
flamingos . . .

I was staying a few days at Kakuru specially to see the marvellous sight
of the flamingos on the salt lake nearby. No description I had ever heard
gave me any idea of what it really was like. The entire shores of this lake for
several miles were absolutely encrusted, as it were, with a moving mass of
thousands upon thousands of these remarkable birds. I went up close to
where they began, and from where, as far as the eye could reach, the entire
shores were surrounded by them, which, viewed from afar, looked like a
monster necklace of pink coral. But they were moving slowly on, all in the
same direction, feasting as they went apparently on a kind of sand-hopper.

When she returned from the wilder parts of Tanganyika to
Nairobi she sought diagnosis of an illness from which she had
been suffering for months. She was, it now appeared, a victim
not only of recurring malaria but of hookworm. 'Disgusting but
easily cured,' she remarks, and spent two days in a nursing home
before travelling on. Other ills were less amenable to treatment.
'I took the opportunity to mention to Dr Anderson that, though
I was now nearly 74 years of age, I was still having those flushings
which for more than 17 years had never left me day or night,' she
noted. 'But all the help or sympathy I got from him was "Most
unusual! It shows that the ovaries are still active." In fact, he
was like all the opposite sex who, just because a woman ate an
apple when she was told not to, seem to consider that as a natural
result all women should have to suffer sexually till the end of
time.'
April 15, 1937 saw her aboard ship and homeward bound
again, to an England which, the ship's radio announced, was
suffering snow, sleet and a cyclone. By August she was off again,
sailing for the United States, Virginia and her nephews. Lee was
– Miss Fountaine really did not approve – married.

Frances was difficult; I never could get more out of her, than 'Ha Ha!',
'No Ma'am', 'Yes Ma'am', for which I could only account from her extreme
youth, only just 18. She is very small and undersized, not particularly pretty.
I could not help thinking that for such a very good-looking and charming
young man as Lee to have tied himself up for life to such a poor little specimen

was scarcely desirable, but he seemed quite satisfied with her. They were living in a little shanty on a few acres of land; one room and nothing else. They scarcely could be said to have a roof over their heads, for on wet nights, Lee said, the rain water soaked through and dripped right on to their bed, and the wind penetrated the cracks in the uncarpeted boards and bare walls. Mollie seemed to think this was quite good enough for Lee; but Lee on his father's side was descended from ancestors who had never been accustomed to such conditions. So I wrote to Rising (the accountant) and told him Lee must have the £500, which he inherited from money of Rachel's, sent out to him at once to buy a farm.

Often on Sunday the boys would both come down to spend the day with me, going to the morning service, then lunching with me at O'Gara's Hotel, while in the afternoon we might drive out in Lee's car. We used to have great games of pool in one of the saloons in Covington, caring not at all that it was not considered quite the thing for a woman to go there, and I soon got to play well enough to be a fairly good match for Lee.

I wished I could have done more for the boys, but my own income is so considerably reduced now with the outrageous income tax and deficient dividends that I really am doing more than I could readily afford. We had many long drives through the magnificent scenery of the Allegheny Mountains, more beautiful than ever now with the autumn foliage, purple blending to vivid crimson of the oaks, with the deep bright yellow of the maples, to find a farm, and at last found 175 acres with a six-room house and another very small house some little distance away; well wooded and more important well watered, with quite a lot of low land under pasture; and both Lee and Frances were delighted with the place.

> She fled south from the increasing cold. Her ship sailed from New York, now the home of her godchild Iris and the girl's father, her old friend Kollmorgen; the romantic Kollmorgen who had once proposed such dubious expeditions forty years before; who had made such improper declarations about her kissable mouth as he escorted her to her tramcar long ago in Vienna. Now the white-haired Kollmorgen bade the old lady farewell with a warm kiss on those lips and 'I turned away somewhat abruptly, feeling too emotional to trust myself any longer; alas for those not-forgotten memories'. It was, she realised, their last farewell.
>
> Miss Fountaine thought Trinidad the most beautiful of the islands of the West Indies, though confessing that she might be influenced by its possession of almost twice as many species of

butterfly as there were in the whole of Europe. To Trinidad, then, she sailed, rejoicing in its luscious warmth. The painful indigestion which had seemed to affect her in the cold quite disappeared, and she felt well again. The pages of her diary now are divided between records of *Preponas* lured to a bait of mixed golden syrup and rum, and accounts of King Edward VIII's abdication in order to marry a commoner, Wallis Simpson. Romantic herself, Miss Fountaine had little patience for the king 'sacrificing his crown and country for a woman in a low class, and an old second-hand piece of goods at that'.

She renewed her friendships at the monastery, St Benedict's, at the hotel nearby and at the agricultural college not far away; there were children, too, to share the less strenuous of her walks. Her 'day' in 1937 she celebrated with a two-hour climb up the Hololo mountain; on the top the air was so pure, she said, that she felt fresher than when she started, and the view over the deep blue of the sea, with the faint outline of the mainland coast in the distance, made the effort worthwhile. After five hours on her feet she returned to her room to begin at once setting the day's captures. She left Trinidad with regret when it was time for her to return again to England.

Oh why does my native land give me such a cold reception? And yet I love it, with all its climatic miseries, and in spite of the feeble health I so soon develop here – every day is an effort, while I scarcely know how to achieve enough energy to accomplish the work I have to do. It seemed incredible that only a few short months ago I had been able to walk at least four or five hours every day, feeling as fit as ever, and now when my day's work was over, I would sometimes wonder how I was going to manage to get back from my studio to the hotel; the indigestion pains which I had never known in Trinidad had all returned, and often I would feel so deadly ill, tired and ill almost all the time.

I had found here in England, that though the Simpson débâcle was freely discussed in the suburbs, everybody being equally agreed in condemning poor Edward, in the West End as represented to me on Sunday afternoons in Skye's drawing-room, the subject was almost tabooed. Even the humorous side of it was not apparently considered open to discussion; all the same several very witty things were being said about him, one of the best being 'that he had given up being first Admiral in the British Navy to become third mate on an American tramp.' To which I added one of my own, which I thought quite good too: 'Esau gave up his birthright for a mess of pottage, whereas Edward VIII had given up his for a tart'. But it was not long before the whole thing became a back number. And the summer passed away as usual, for the most part cold and wretched.

I now made plans for going to Indo-China, having decided to go by P & O to Singapore, and from thence on the Messageries Maritimes to and from Saigon. So I said goodbye to my friends at the British Museum, telling Mr Gabriel that it would be a long time before I should be bothering him again over the classification of my specimens. At the end of August the short English summer was over, and September was cold and wretched, so that as usual I was longing to get away, hoping that I should then feel quite well again, which I did.

At Singapore we visited the Air Port of Imperial Airways, as I was hoping there might be an air service direct between Singapore and Saigon, but alas there was not, so I decided to leave on a rather small boat about ten days after my arrival. Meanwhile I was introduced to the French Consul, who provided me with a letter of introduction to the Gouverneur de Cochinchina in Saigon; when I had trouble with the French officials who came on board just before we reached Saigon, seemingly bent on annoying me as much as possible, the letter worked wonders.

Just as we were arriving up against the wharf in Saigon, a thunderstorm of exceptional violence was breaking overhead, and very soon torrential rain and wild bursts of thunder, with vivid and almost uninterrupted flashes of lightning, were doing their utmost to make this landing even more difficult than I had anticipated. During a slight abatement in the storm, I decided to land somehow with my smaller luggage, but the douane was some little distance away, and as there was no roof to protect me and the road was deep in water with torrential rain still falling, I was soaking wet from head to foot long before I got there.

The business of the douane was soon finished, but there are no taxis in Saigon, and the storm was raging now more fiercely than ever; thus it was that with thunder shrieking and screaming overhead, and with vivid flashes of lightning cracking and hissing through that most terrific downpour of rain, I started off in a rickshaw with my suitcases and other small belongings in another rickshaw, till at last I arrived at the Continental Hotel, half drowned and more than half stupefied. However, a nice large room, with private bath and toilet and a rapid change into dry raiment, went a good way towards reviving my drooping spirits.

My first move was to find a sufficiently good and not too expensive second-hand automobile. I eventually purchased a second-hand Citroën car for 700 piastres, equivalent to about £50. I also found a native chauffeur; a rather old man. I got the hotel to ring up the governor's secretary and was told I should go to the Residency to present my letter of introduction that same afternoon at 5 pm.

I think my old chauffeur was considerably impressed when he drew up under the portico of a very palatial edifice. I was admitted with all due pomp and shown into a large empty room on the ground floor. I waited and waited, and was beginning to feel rather depressed, wondering if perhaps my prestige with the old chauffeur was going to suffer rather than be enhanced, when I was taken up a long flight of stairs and ushered into the presence of a very pleasant young Frenchman who explained there had been a mistake and that Monsieur le Gouverneur would be 'très occupé'.

He was so good looking, this young Frenchman, and so charming in his manner that I could not feel really annoyed, and he insisted on my retaining the letter, which (I reflected) would be much more useful to me to show as occasion might demand, than if I had merely made the acquaintance of the Gouverneur during a short period of conversation which I should probably not have enjoyed nearly as much as I did with that good-looking young Frenchman. I only trusted that my chauffeur waiting patiently outside would never know what had been passing within those awe-inspiring precincts, otherwise my prestige would be damaged irreparably!

My first objective when I left Saigon was Phanthiet in the province of Annam, where I had been told I should find forests at no very great distance, and then to Dalet and over the mountains to Djiring. No doubt the old man really was an expert driver, but he insisted on stopping at frequent intervals to tinker with the wheels of my car, changing them from one to another, and then replacing them where they had originally come from, which I could not but think was a most unnecessary proceeding. At other times he would tinker with the engine. I could do nothing but wait impatiently, or wander off a little bit down the road, as he generally vouchsafed no answer to any remark I might happen to make on the subject; and if he did, it was in a most rude, insulting manner. On one occasion he seized an opportunity to replenish the car with water, for which he promptly commandeered a Chinaman's hat, adapting it admirably to the occasion, though it would have made less mess had it not been for that hole in the centre of the capacious crown.

Dalet was a mountain resort, and awfully cold, and I had not come to Indo-China to suffer from the cold; neither did I anticipate much collecting to be done here. The car had to go to a garage to be overhauled, my chauffeur's tinkering having done more harm than good. On the way back to Saigon the car broke down again, and then ran out of petrol; while the chauffeur went to find some, I closed the windows and wondered if I should see a tiger or any other wild beast of the forest. I didn't like it at all, sitting there alone and helpless in the pitch darkness until help arrived. But when, hours later, I entered the brilliantly lighted dining saloon in the Continental Hotel in Saigon and heard the band playing, the horrors of that long, wearisome day were soothed away.

I spent Sunday at Pnom-Penh, the capital of Cambodge, a very clean, well laid-out little town, with wide streets and lovely little gardens skilfully laid out, and went on to Angkor. I had heard a good deal about the wonderful ruins here, but no one had ever conveyed to me any idea of what they were really like; I had seen nothing at all like them in any other part of the world, and I have poked about a bit. Not only the Angkor Vat, but right away in the deep jungles, where my work led me to go, I would come across old relics of temples, and sometimes whole villages crumbling away and overgrown with the dense tropical vegetation, and always decorated with carvings of the most marvellous artistic merit. And where are now the hands that sculptured those old stones? Dead and forgotten long ago, a race that has vanished.

There was good entomological work too to be done in the neighbourhood of Angkor, and many fine collecting places, and though I constantly met with opposition from the old chauffeur whenever I desired to penetrate a little further than usual into the jungle, I made him come along. I believe he was in constant fear of meeting a tiger, or some other wild beast; after a time he began to see that there was nothing more formidable than troops of monkeys to be encountered in those forests, at least in the daytime. In the end the old man conceived an unaccountable and insatiable love for moths: 'Un papillon de nuit!' he would cry out, and even tigers were forgotten in the excitement of the moment. I had a good many pupae, some of which were *Ornithoptera*, which had been brought to me by some small boys, who had of course been suitably rewarded.

Christmas came and went (almost unremarked here, with a pagan population of French and natives) and New Year's Day. The long drought was now in full progress, so I thought I would go back to Saigon, return to Singapore, and pay a second visit to Indo-China for the wet season, during which I felt sure I should be able to do very well indeed. Had the old chauffeur been less insolent, less scared in the jungle, and less prone to smoke cigarettes and imbibe tea or coffee and eat disgusting food all day long in my car, in fact had he not been altogether thoroughly objectionable, I might have felt some regret at parting from him, whereas I was thankful to get rid of him.

In Singapore again I was welcomed by Captain and Mrs Hancock; I would lunch with them, and play chess with Captain Hancock. He was rather annoyed at first at my determination to return to Indo-China, where

he seemed to think I should lay my bones; so I promised if I died I would
write and tell him at once; adding that as I had come out here specially to
work Indo-China, I was going through with it. And I was much encouraged
in this resolve by a letter from Captain Riley, in response to a consignment
of insects I had sent to him, saying he had looked them over and seen many
things which would be most useful to the British Museum.

When I returned to Saigon, where my car had been taken care of, I
secured another chauffeur, a younger man. He drove out to the hills near
Nha-trang, then to Tourane; and on to Vinh. It was cold and rainy; and the
rice fields looked desolate. The country roads seemed overflowing with
natives, their big hats looking like monster mushrooms. They were all carrying heavy weights, generally by means of a bamboo pole slung across their
shoulders, the women being by no means exempted. Sometimes two good-
sized straw stacks would be seen approaching, the slender legs of a Chinaman staggering in between; there were buffaloes, great coarse, ugly brutes,
generally gorging on the grass at the edge of the rice fields with a small boy
sprawled about on the broad plateau of their backs, which in the rain looked
very slippery and most uncomfortable.

The French do know how to lay out a town, and Hanoi was no exception;
wide boulevards and well-kept gardens, a blaze of colour with gorgeous
flowers. Here I was found a new and first-rate chauffeur, Doey, and one of
the best boys I have ever had to help me in my work: Huong, a pale rather
sickly looking child, fatherless, and very very poor, living with his widowed
mother and three or four little brothers younger than himself. I was glad to
take him on, and soon found that he caught on amazingly to this work.

With these two helpers, Miss Fountaine was to have one of
her most enjoyable and successful seasons. She records her 'day'
again, 15 April 1938; how they drove and tramped through the
countryside, finding food plants, capturing – to the boy's delight
– two species of *Zetides* and a long-tailed black-and-white
Pathysa on the wet, muddy road with one sweep of the net; how
she slipped and fell on a muddy roadside bank and little Huong
didn't laugh as most small boys would have done but rushed to
help her, anxious till she told him all was well. The boy found
some butterfly eggs, and food plants that her hungry caterpillars
needed desperately; she watched them feed, sent the boy and
her driver to eat and lunched carefully herself – the Hanoi area
was cold and those indigestion pains had returned. 'I sometimes
wonder if these pains are not something more serious than indigestion – but could I be my mother's daughter and not imagine

I am going to be the victim of every disease to which the human being is prone?' She went to bed very happily.

It was a big storm, after several days of intense heat. The evening sky was dense with threatening clouds, and the ever-increasing peals of thunder and flashes were going to make the night terrible. I fear these tropical storms now, and I was hastening to undress and get into bed, and sleep if possible till it was over. The rain was coming down in torrents, and the thunder and lightning were almost unceasing; then came a flash of fearfully vivid lightning, a deafening roar of thunder, and a tremendous crash, all in the space of a second, instantly followed by dense darkness, for the electric light was in that one moment dead. I was only half undressed, so there was nothing for me to do but sit quite still in my darkened room and wait until one of the hotel boys brought me up a lighted taper, by the feeble light of which I was able to complete my toilet, and slip into bed beneath the mosquito net; while still the thunder roared and shrieked. Next morning the manager told me that the hotel had been struck by lightning.

After this there was plenty of rain, so that Huong and I had to keep not too far from Doey and the car; but we were doing well with our work; Huong had found the ova of one of the larger *Charaxes* on the leaves of a large shrub, and I soon had a big family of these interesting larvae, feeding amongst many others on the verandah outside my room. Besides having caught on to the work at once, Huong was such a happy, contented little fellow, seeming really to enjoy our long treks, only a bit crestfallen on the days when he had not found all he had hoped to find, and I got to love the little white figure running after me along those mountain paths. What was best of all, the boy was really getting stronger and better in his health. It was a delightful life I was leading now, absolutely to my liking, for I was very busy, and the French were charming whenever I met them.

The ricefields which were dry stubble when I came to Hagiang more than two months ago, now lay like huge mirrors, reflecting the mountains and the sky with young shoots of rice just beginning to come up. I had done remarkably well with my collecting; no less so did I find it now at Tuyen-Quang. Huong had soon become clever at identifying food plants, and it was most amusing to watch the superior airs he would put on with Doey, who also was getting quite keen on this work but was a most good-natured person, and quite enjoyed being patronised by his 'petit frère', which he said he considered Huong. No doubt the wet season is the best for insects, noxious and otherwise, and for working out the early stages, but one can have too much of a good thing, and the heavy rains up in the mountain caused the Riviere Claire to become a turbid torrent, increasing daily till at last it began to overflow its banks, and the road in front of the Hotel des Mines was flooded. At first this was only enough to be the delight of all the younger members of the community, but the water rose rapidly till soon there was no way of leaving from the front of the hotel, except by boats. My bedroom was on the second floor, so I was in no personal danger of being drowned, but the back yard was also about fifteen feet deep in water; how was I to get to the jungle to procure the necessary food plants?

I was sorry to leave, and the five months I had spent there will always remain one of the happiest memories of my old age. After a short stay in Hue because of precautions against cholera, we were off and away down the coast. The never-ceasing delight of Huong at all the new experiences he was having was a joy to me too; his first sight of the sea was a revelation. Doey too had all the keenness of a very young man in spite of a little retinue of sons left behind in Hanoi. One evening when it was getting dusk and we were still some distance from Phanthiet, a peacock crossed the road in front of the car. Huong had never previously seen a peacock, and before I could realise what was happening Doey had pulled up and boy-like Huong was running as hard as he could go, back along the road, evidently with the intention of turning into the forest, in hot pursuit of the peacock. I did not appreciate this delay, so I promptly told Doey to call after him and tell him that now it was getting dusk he ran a considerable risk of meeting a tiger – big game hunters often came here to shoot them. This information brought the boy back with amazing rapidity, and I soon realised that Doey was driving over a not specially good road, at a pace I should never have thought the old Citroën capable of, and this pace was kept up till we were clear of the forest.

As we approached Saigon the next day, I think I was infected with some of Huong's excitement. It was no easy matter in the complicated maze of streets to find the Continental Hotel; Doey and I were quite at a loss. During the week I stayed here, we made several excursions, and there was always the same difficulty in getting back to the Hotel, and I soon discovered that Doey had no better bump for locality than I have; indeed it was generally Huong who pointed out the various turns we ought to take, at once assuming that air of superiority which never failed to appeal to Doey's sense of humour.

I returned to the area around Angkor, but the best localities were now under water and unapproachable. My time in Indo-China now was fast drawing to a close. I stayed on a month in Saigon doing quite a lot of good 'business' in the forests at the Trian Rapids and other places on the road to Phanthiet. As the time approached for my departure, I felt more and more sorry to go; not a little on account of Doey and Huong, both of whom seemed really sorry too. One day when I saw them talking very seriously together, I asked Doey what he was saying to Huong to make him look so sad, and it seemed he was telling him that neither of them would ever again have such a good padrone. Well, I had been good to them, but they on their part had given me excellent service, and I was just as sorry to part with them. Indeed, had I not now almost filled all the store boxes I had brought with me I don't believe I should have left Indo-China for some time still, for I knew I might never be so happy again.

Doey and Huong were both perfect to the very end; they were leaving that same night by train for Hanoi, but each of them had given me his home address, and I promised them if ever I returned to Indo-China, they should come back, if they wished, to work with me again. I felt sorry indeed when, standing on the upper deck of the ship, I looked down and saw them all drive away, Doey at the wheel as usual with Huong sitting next him and Monsieur Goubert, who was buying the car. And so that's that, I thought, and turned sorrowfully away.

> Miss Fountaine spent that Christmas of 1938 in Singapore as the guest of the Hancocks, a family with whom she had become acquainted; one photograph in the diaries shows her in a family group beneath draped paperchains; another shows her gaunt-faced but smiling, wearing a Christmas cracker hat: 'the joy is with me still,' she wrote some months later. After Christmas she paid a visit to Celebes, first hiring a Dutch girl as her collecting assistant but soon discarding her; not only was she lazy, but she went into shrieks of laughter on being shown a hundred eggs of a rare butterfly with which the triumphant collector had returned. 'I ought to have known that mine was not woman's work,' declares Miss Fountaine scornfully. Instead, she took to travelling and sharing expenses with Mrs Proctor, a young woman who was collecting material for a book on the island; they became friends.

Though she was one of those women who are possessed of an unusual attraction for the opposite sex we got on fine together, and I never liked her better than when after some time of animated conversation, she would suddenly say: 'Now shut up! I want to concentrate. Go out and look for some of your caterpillars.' And this I would do, but always with very poor results, as the country through which we passed, though remarkably interesting from other points of view, was practically denuded of forest, and almost all under cultivation, with the eternal rice-fields in every available spot.

Once really in the Taradja County the type of natives, and more especially their houses, was quite different, and there were colossal cliffs up the sides of the mountains with tombs in caves and effigies of the deceased ranged along a narrow ledge outside, hundreds of feet up the sheer sides, one of the most remarkable sights I have ever seen. Altogether this little jaunt with Mrs Proctor was quite delightful, and if she did get drunk the last evening we were together, she was just as charming drunk or sober, and I had no reason to complain of anything.

By the time Miss Fountaine's 'day' for 1939 came round she was back with her friends the Hancocks in Singapore; it was good to wake up in their house on a fine morning with misty clouds hanging over the city. It was a world, she suspected, going down into war; she looked at the prospect with anguish, but her own cure, she thought, could not be far away: 'I do not think I shall live much longer; perhaps this is the last time my worn out brain will be called upon to conjure up the past, and once more accomplish the task I have set myself now for more than sixty years.'

But there were still butterflies; a *Polites* pupa had just emerged. There were her hostess's children; how fine to watch their fearless diving at the swimming club, how warming to have a little girl sitting at her elbow to watch the careful setting of a butterfly, just as her nephew Lee used to do. There was a letter from her young friend in the Celebes: 'many admirers, cock-tail parties and dancing all night, and Hudge apparently being brought once more to heel . . . it was nice of her to say she had followed and profited by the advice I had given her in this affair of the heart'. There were preparations for a wedding party; how lovely to see so much happiness. There were visitors arriving from New Zealand, with whom she could talk about her visit there with dear Charles.

It was after ten o'clock when Miss Fountaine retired, 'but so loth am I to end yet another of these happy days I would gladly have stayed up even longer'. There were last-minute arrangements to be made, for she was to start early next morning. There was the record of 15 April 1939 to be completed: 'And the last I knew of this day was to hear the clock strike the hour of midnight.'

She copied out this entry three months later in England, adding the date at the end in the usual fashion – July 10: 1939. It was page 3,203 of the diaries. Then she wrote, in the same still perfectly legible hand, 'April 15: 1940' ready for next year, with two lines scored above and a line below.

The rest of the volume is blank.

THIRTEEN

After the Silence – Envoy

Margaret Fountaine had returned to England, but not to stay. The fact that she was nearing her 78th birthday would not have been reason enough to sit at home shivering over a fire; not while there was tropical warmth to be found, not while the butterfly collection needed further specimens, not while there were butterflies as splendid as flowers in flight to be captured – or better still to be reared from egg or caterpillar, the chosen specimens kept but the majority set free.

War had broken out that autumn and sea voyages were not to be lightly undertaken. On land there might be a deceptive calm between the armies, but at sea the sinkings had begun. All the same, she sailed for the West Indies, for Trinidad, where eight years earlier Father Maurus of the monastery at Mount St Benedict had been such an enthusiastic companion in her collecting expeditions, to stay at the monastery guest house, to know once more the benison of warmth, to see again how beautiful the world could be.

Before she left England there had been what must by now have become annual formalities to be completed. Her diaries must be repacked in their black-japanned metal box. She was now into the twelfth volume, her handwriting in old age no less clear than when she had begun the first as a girl of 16. In with them went the little blocks of carbolic with which, as an entomologist, she was familiar. The blocks kept corruption from the fragile beauty of butterflies; they might keep worm and moth from that long written record of her life. If she should not return from this journey the manuscripts would remain locked away for many years. That had been provided for in her will, made two years before.

In it she bequeathed to the Castle Museum at Norwich 'all

my collection of Diurnal Lepidoptera and other natural history specimens . . . together with all cabinets containing my collection and the tin box and contents (which are in fact books in manuscript and papers) . . . the said collection shall be kept as a whole . . . and be known for all time as "The Fountaine–Neimy Collection".'

She had been faithful to Khalil Neimy, to the memory of the young man whom she had at first despised; who had wooed her so passionately and in such uncertain American-accented English and won her even though he was so plainly what Mamma and her friends would have described as 'impossible'; won her despite herself and despite what even she admitted was his mysterious and uncertain domestic background; she had learned at last to trust and accept his devotion. He may at first have pursued diurnal lepidoptera only as a means to her favour, but he came in time to be a skilled, determined and to an extent knowledgeable collector, and in this at least her equal, and Miss Fountaine declared it to the world. The collection was his work as well as hers and must carry his name for all time. It carries his name today.

The collection was left to the Castle Museum with a small bequest, £300, to pay for its care, and with a further condition: the black-japanned tin box was not to be opened before 15 April 1978. No explanation was given for this. Not until the box, and the diaries, were opened in 1978 was the reason discovered. That day in 1939 the books were packed away. With the twelve volumes was a letter:

Before presenting this, the story of my life, to those, whoever they may be, one hundred years from the date on which it was first commenced to be be written, April 15 1878, I feel it incumbent upon me to offer some sort of apology for much that is recorded therein, especially during the first few years when (as I was barely 16 at the time it was begun) I naturally passed through a rather profitless and foolish period of life such as was, and no doubt is still, prevalent amongst very young girls, though perhaps more so then, a hundred years ago when the education of women was so shamelessly neglected, leaving the unitiated female to commence life with all the yearnings of nature quite unexplained to her and the follies and foibles of youth only too ready to enter the hitherto unoccupied and possibly imaginative brain.

The greatest passion, and perhaps the most noble love of my life was no doubt for Septimus Hewson, and the blow I received from his heartless conduct left a scar upon my heart which no length of time ever quite effaced.

For Charles Neimy, whose love and friendship for me endured for a period of no less than 27 years, ending only with his death, I felt a deep devotion and true affection; and certainly the most interesting part of my life was spent with him, the dear companion, the constant and untiring friend and assistant in our entomological work, travelling as we did together over all the loveliest, the wildest and often the loneliest places of this most beautiful Earth, while the roving spirit and love of the wilderness drew us closely together in a bond of union in spite of our widely different spheres of life, race and individuality, in a way that was often quite inexplicable to most of those who knew us.

To the reader, maybe yet unborn, I leave this record of the wild and fearless life of one of the 'South Acre children' who never grew up and who enjoyed greatly and suffered much.   M. E. Fountaine.

> The letter was placed on top of the diaries, the box shut and padlocked. Miss Fountaine drew a canvas cover round it, strapped the cover tight. To the outside she attached a warning: 'Not to be opened until April 15, 1978.' She signed and dated the label: 'September 5, 1939.'

Seven months later, on 21 April 1940, Miss Fountaine was discovered by Brother Bruno, one of the monks of the Mount St Benedict monastery on the island of Trinidad, collapsed by the roadside. Her butterfly net was nearby. She was so plainly seriously ill that, despite her protests and struggles – she was a strong woman, he recalls – Brother Bruno picked her up and carried her in his arms to the guest house near the monastery, where she had been staying. She had suffered a stroke, and died very soon afterwards. She was buried next day; this was the tropics.

In due course her will was read, and the 20,000 butterflies in their handsome mahogany cases were despatched across a blacked-out and bomb-scarred wartime England to the Castle Museum in Norwich, along with the black-japanned box. As the will asked, Mr Norman Riley of the British Museum – the young Captain Riley who had been so helpful to her on her visits to the Museum as long ago as 1913 – supervised the moving of the collection, getting £100 for his trouble. Her sketchbooks, the meticulous paintings of butterfly eggs, caterpillars, pupae and food plants, went as she had wished to the Natural History

Museum in London, her collecting equipment and books to the youngest member of the Royal Entomological Society. There was a legacy of £100 to Angelina, Italian friend of the days when they were both girls and cycled together through the streets of Milan. Her jewellery went to her God-daughter Hildegarde . . . daughter of old Kollmorgan who had so improperly admired Margaret a lifetime ago in pre-war – in pre-first-war – Vienna. If Hildegarde should die first then the jewellery should go to Frances, wife of Margaret's American nephew Lee. Frances, it seems, had been forgiven for being such a poor match for handsome Lee. The rest of her estate, her house at 100A Fellows Road, Hampstead, her pictures and other effects, were left to Lee and his brother Melville.

The various conditions of the will were carried out. The black japanned box lay at the Castle Museum for nearly forty years. In April 1978, two days late (because 15 April that year fell on a Saturday) the box was opened. I have described elsewhere that opening; enough now to say that in 1980 extracts from the earlier diaries were published as *Love Among the Butterflies*. Even before the book appeared, however, newspaper accounts of the finding of the diaries had begun to help fill in more details. It was, for example, probably some time in that last summer that Miss Fountaine had paid a visit to a photographer, Barbara Ker-Seymer, at her studio in Bond Street. It was still her habit to use a portrait photograph as a frontispiece to the diaries. If there had ever been a measure of vanity in this practice, there was no longer; the portraits had become a cruel demonstration of Time's penalties. They showed an old woman's lined, gaunt and tired face; but, then, Miss Fountaine's portraits never did justice to her spirit, nor indeed to her indisputable charm.

Miss Ker-Seymer did not know her subject, but Miss Fountaine explained that they were related; Margaret's aunt, Caroline Fountaine, had married Sir John Lawes, the millionaire scientist and industrialist who was Miss Ker-Seymer's great-grandfather; the Sir John Lawes whose financial skill had nurtured the family bequest which gave Margaret her freedom to travel, which had made the whole lifelong butterfly-collecting adventure possible. Among Miss Fountaine's multitude of cousins was Lawes's daughter Skye, author, painter, society hostess and redoubtable 95-year-old, and Skye had passed on to her old friend the name of her young kinswoman. Miss Fountaine very probably welcomed an opportunity to put some work in the way of a woman in a profession where women were not common.

Forty or more years later, Miss Ker-Seymer remembered Miss Fountaine disappearing into the studio changing room, to emerge soon afterwards in the then somewhat startling costume she favoured for butterfly-collecting – a man's checked cotton shirt and a striped cotton skirt, both with additional pockets sewn on.

'She was wearing cotton gloves with the tips of the first finger and thumb cut off. A heavy black chain with a compass on the end was attached to one buttonhole of her shirt, and she was wearing a cork sun helmet. She wouldn't hear of a studio portrait but was very particular about how she was to be photographed. She sat down bolt upright clasping a large butterfly net and a black tin box for specimens . . .' Miss Ker-Seymer never saw her again.

Another glimpse of Margaret Fountaine in her later years came from Mr Norman Riley, now retired from the British Museum where he had so often been her adviser on the finer points of the identification of rare butterflies. In response to my request he wrote the following recollections of her, beginning, he said, when she was announced one summer afternoon in 1913, at the Natural History Museum in Kensington:

'Having heard something of Miss Fountaine's exploits the announcement conjured up visions of a well-worn battle axe . . . instead I met a tall, attractive, rather frail-looking, diffident but determined middle-aged woman. The strongest impression she gave me was one of great sadness. It was not long, however, before I discovered that this veil of sadness could be penetrated by self-deprecating flashes of humour that quite transformed her . . .'

Miss Fountaine's visits quickly developed a regular pattern. 'She would appear silently and usually unannounced and produce a box in which she had pinned several choice specimens she had not been able to identify satisfactorily, often rarities not illustrated in the standard reference book. A considerable search was often involved; it did not matter, she always seemed perfectly content to wait patiently until I had the answers. Then with gentle and obviously sincere thanks she would fade away until the need for further help arose.

'In later years, her unfamiliarity with London and its traffic led her at times to seek the company of my wife when a shopping expedition became necessary, usually just before her departure to some remote part of the world. One of these shopping expeditions was opened by the purchase of six pairs of plimsolls. When

Edith questioned the suitability of such footwear in the leech-infested jungle to which we knew she was going, she was told that, on the contrary, they were ideal, being light and comfortable and disposable – and as to the leeches, she took a creosote bath once a week and they never came near her! On another occasion, before leaving for India, where she was threatened with the possibility of having to stay at the viceregal palace, she decided her wardrobe was hardly suitable for such surroundings. Having bought six very lovely and expensive pure silk dresses, she turned to the seamstress and asked her to open the seams of the skirts and insert a large pocket on each side to carry her butterfly boxes when she was in the jungle – it would be wasteful not to use up the dresses.

'Of her Hampstead studio, where we several times visited her, I remember little except a range of handsome mahogany butterfly cabinets, and a grand piano which at times she would play quite enchantingly. My son, already a keen bug-hunter, greatly admired the cabinets and, child-like, let it be known that he wished he had one. Within a fortnight he had one, delivered to the door as a present from Miss Fountaine, an act of generosity very typical of her.

'Talk of bug-hunting led my son to mention Halnaker Gallop, near Goodwood, as a good spot. Miss Fountaine had ridden there as a girl. As she had not seen it since then, we invited her to spend a weekend with us at Selsey, not far away. We went to the gallops, and, as she says in her diary, she enjoyed watching our two 'adorable children' chasing Meadow Browns in the long grass. What she does not say is that, when half a mile from the car, we were hit by a rain squall that drenched us to the skin. On getting home the children were rapidly put through a hot bath. On looking for Miss Fountaine we were astonished to find her enveloped in clouds of steam, sitting on a chair almost on the open log fire. She said she had no need of a bath and would be perfectly all right as soon as she had dried out. And so she was.

'Talking of baths, she recalled that once in Queensland she asked for a bath after a long hot ride and was ushered into a screened area under a large tree. As everything she needed appeared to be there she stripped, only to find that there was no water. Her shouts for help were answered by a voice from an overhanging branch of the tree, saying "Coming, Miss", followed by a bucket full of water tipped straight over her.'

From a lifetime of acquaintance with naturalists, Mr Riley summed up Margaret Fountaine's place among them: 'Naturalists, particularly the kind who take a pride in building up valuable collections, are notoriously bad at recording their observations for the benefit of others. Miss Fountaine was one such. Her knowledge of the ways of tropical butterflies was profound, probably unique, but only here and there in her diaries and published accounts of her journeys do unexpected records of odd events crop up: at Poona a small bird persistently flew in through an open window and stole her caterpillars from under her nose; in Durban another small bird was seen fleeing from the deliberate attacks of a large Swallow Tail butterfly.

'She was rather a loner, with few really close friends even among entomologists and this and her uncontrolled wanderlust may account for her recording so little of her work in print or in any systematic form. She undoubtedly enjoyed her life of butterfly catching but failed to catch that one deep happiness she ever sought.'

It was while working on *Love Among the Butterflies* that I made the acquaintance of Mel Fountaine, Margaret's nephew. It had occurred to me that though most of those named in the diaries would inevitably no longer be alive, her two nephews Melville and Lee, whose childhood she records, would be no more than in their sixties and might be traceable, even among two hundred and fifty million very mobile Americans. I consulted telephone books for Virginia and West Virginia, where Miss Fountaine had last met them, and found the name of Fountaine did not occur in unreasonably large numbers. As a long shot, I asked the New York office of the *Sunday Times* to telephone them all and inquire of each Fountaine if they had had an English Aunt Margaret who died in 1940.

Some days passed. The New York office replied that a dozen or so Fountaines had been tried without success; there was only one Fountaine left and that number, in the small town of Staunton, Virginia, was not replying. I asked New York to have one last try. The next day the New York office rang back; they had located the Fountaines in Staunton, and yes, they had had an English Aunt Margaret, but if I wanted to speak with them I'd better hurry – they were going on vacation next day. I telephoned the Staunton number. Yes, indeed, they were just off on vacation – to England.

We met at their holiday hotel in London; Charles Melville

Fountaine and his wife Elnora were on a fortnight's tour of historic places. He remembered his aunt well — no man who as a shy teenager has had to accompany an elderly, autocratic and highly competent billiards-playing aunt into a small-town American pool hall to the amazement of the citizenry is likely to forget the experience, any more than the same boy would forget the aunt generous enough to buy an old car which he might drive. But Mel knew little of Miss Fountaine's — and his own — family background. He knew she had lived in Bath, but nothing of the family home in Norfolk nor the six-centuries-long Fountaine pedigree, and he believed her to have been the last of the English Fountaines.

I showed Mr and Mrs Fountaine some of Margaret's diaries, and persuaded them to play hookey for a day from their tour. We drove to Norfolk to discover their English kin and family roots: the little church in the hamlet of South Acre where Mel's grandfather, the Reverend John Fountaine, had preached; the old rectory across the road where Mel's father, and Margaret, had been born; the seventeenth-century Narford Hall not far away, the family seat where Andrew Fountaine, head of the family in England, now lives. Later we were to go on to the village of Salle where still earlier Fountaines, ancestors of Margaret and of Mel Fountaine, are buried — along with their neighbours the Boleyns, one of whose girls had the misfortune to marry Henry VIII — and to Rochester Cathedral where hang the coat of arms of another ancestor on dear Mamma's side of the family, a seventeenth-century Bishop Lee Warner.

Publication of *Love Among the Butterflies* brought me more glimpses of Miss Fountaine. From Wincombe in Dorset Mrs G. M. Weall wrote that she was the daughter of Miss Fountaine's one-time landlord, Commander John Smith-Wright, and recalled that '. . . about 1925, Miss Fountaine rented the large studio at 100 Fellows Road in Hampstead which was our home. To us children she seemed a very old lady; her hair was grey by then as she was in her mid-sixties. We used to be invited in to see the butterflies, and one year Miss Fountaine sent us back a grey African parrot from her travels. Much to our surprise she came back one year with her fiancé, whom we knew as Mr Neimy. He was very jovial and friendly, and fascinated us children by the number of gold fillings in his teeth . . .

'Some time in 1929 she came back and was waiting for Mr Neimy's arrival; I believe they were going to be married. One day she came down to our nursery looking very distraught and

137

asked my father to come outside. There she told him that Mr
Neimy had died . . . shortly after that we left London and I
think Miss Fountaine took on the lease of the house.'

I received a letter, too, from Felicity, the little girl with whose
family Miss Fountaine was staying in Singapore when she wrote
the last entry in her diary: forty years later, Mrs Felicity Little.
She remembered the 'butterfly lady' and her visits, and gave me
instructions from which I might find the spot from where Miss
Fountaine, leaning from a window of their house, had taken a
photograph looking across Singapore to the cathedral.

The house is no longer there, as I discovered when I revisited
Singapore. A tall tower of flats stands in its place; and the other
towers of Singapore's new prosperity, the hotels and office blocks,
now hide all sight of the cathedral, where she prayed . . . though
the cathedral itself remains, a monument to Victorian piety
carried, like the empire itself, to the ends of the earth. Carried,
and suitably modified; the pews are cane-seated to ease the
impact of tropical heat on worshippers. One of the hymns, its
number still in the frame from the last service, was perhaps an
appropriate reminder for any biographer: 'O God, our help in
ages past'. Watts's words, as the devout will remember, warn
us that

> Time, like an ever-rolling stream
> Bears all its sons away
> They fly forgotten as a dream
> Dies at the opening day.

But perhaps, for one of Time's daughters, recollection may last
a little while longer, while her diaries survive.

And long-lasting bronze. Among the readers of *Love Among
the Butterflies* was Brigid Keenan, a former colleague on the
Sunday Times, then living in Trinidad, from where she wrote
flatteringly about the book and asked if money could not be
raised to set up a plaque to the memory of Margaret Fountaine.
I had to reply that I had raised the same question about the
possibility of a simple memorial to Miss Fountaine in the church
at South Acre, where she had been christened and where, her
will had directed, her ashes were to be buried if she should die
in England. My suggestion had met with a discouraging
response, I explained.

I should have known that Brigid would not be so easily discouraged. Three months later she wrote again. 'I went ahead
and *did* the plaque for Margaret Fountaine!'

More than a little shamed by this, I wrote to ask for details. She replied explaining how she had met Brother Bruno, the monk who had gone to Miss Fountaine's aid when she collapsed by the roadside; he still remembered her death.

'He was sure she died of a heart attack. She was taken down the hill (probably on a mule since there was no decent road then, I believe), and buried in Woodbrook Cemetery in Port of Spain. The High Commissioner's wife and I went and looked for the grave and found it, all overgrown. According to the cemetery records there are several people in her grave, indicating that she was buried at the expense of the city. . . The cemetery is rather romantic, not far from the sea with tall palms waving above, not such a terrible place to be. There is nothing on her grave to indicate that she is there; it was only from cemetery records that we found her.

'I thought it a great shame that such a plucky old girl should have died so far away from home with no companions, and when I discovered she was in an unmarked grave I felt positively sad about it and decided to see if I couldn't raise a little memorial to her.'

In this sadness my correspondent was perhaps unnecessarily perturbed – Miss Fountaine had more than once reflected in the diaries that she would probably end her days in an unmarked grave, and like Byron she had preferred to think of her last rest as being somewhere less chilly than the cold clay of England.

Brigid went on: 'I approached Abbot Hildebrand at Mount St Benedict and he said he would have no objection to a plaque being put up there – it seemed a more romantic spot than the overcrowded cemetery, and after all, she died there.

'I gave a short talk about Miss Fountaine at the UK Women's Club of Trinidad and then passed the hat round, and with that money, plus two large donations from other British women living in Trinidad, we commissioned Ken Morris, a respected and admired local sculptor. He made a square plaque with a drawing of a butterfly on it and an inscription chiselled out by hand:

In memory of Margaret Fountaine,
English traveller and butterfly collector.
Born 1862, died at Mount St Benedict, 1940.

'We put this up on a stone overlooking the most stupendous view over the Caroni plain, on the edge of the cliff not far from the guest house where Miss Fountaine died, a gorgeous and romantic spot.

'The only part of the ceremony that didn't go smoothly was before it started, when I found that the miles of electric cable I had brought up (so that Ken Morris could drill holes for the screws holding the plaque to the stone) did not reach the nearest electric plug. I had a vision of the ceremony turning to farce – my having to stick the plaque on the stone with chewing gum, not that I had any of that either, and coming back to fix it at dead of night. But Ken Morris valiantly made the holes with a hammer and chisel while I leaned over the fence watching the High Commissioner's car winding its long way up the hill, biting my nails and wondering if I should throw tintacks over to give them a puncture. However, the last screw went in as the limousine swished up, so all was well.

'The ceremony was very touching. The British High Commissioner, David Lane, made a short rather moving speech about Margaret Fountaine. Then Abbot Hildebrand said a prayer for her (I was worried that he would frown on her morals because of Khalil, but he said: "Which of us is without weakness?"), and Brother Bruno beamed on. The local newspapers reported it and the UK Women's Club was represented. Afterwards we all had bread and honey (from the monks' own bees) and home made cakes and tea in the guest house . . .'

In whatever butterfly-filled Paradise she now inhabits, Margaret Fountaine would have appreciated that. It makes a good end to her long story.

One puzzle, however, remains. In the September 1940 issue of the *Entomologist*, Mr W. G. Sheldon, a distinguished colleague of Miss Fountaine, wrote a brief obituary. In it he said: 'For some time her health had been affected by shock sustained through an attack by a submarine on the ship on which she travelled across the Atlantic last autumn.'

While working on *Love Among the Butterflies* I endeavoured to find out more about this last adventure; for one thing, having read the diaries through, I could not see Margaret Fountaine affected by such a şhock. As Mr Sheldon had said in the same obituary: 'She was absolutely fearless . . . oblivious to the presence of any inhabitants of tropical jungles, be they lions, tigers, leopards, poisonous snakes, or malaria, yellow fever and other tropical diseases. . .' I reckoned that a submarine attack would have done no more than stimulate her to heights of patriotic defiance. But what was this ship on which she had travelled, and where and when was it attacked? Mr Sheldon was dead; Mr Norman Riley, who might have known (a sick man when he wrote his account of Miss Fountaine for me), had died very shortly afterwards. Checks among the files of the National Maritime Museum and the Imperial War Museum failed to find a likely vessel. Inquiries in Trinidad were equally fruitless. Names of various ships were followed up, passenger lists checked, in vain.

Perhaps some clue to this last of Miss Fountaine's adventures will one day appear. On the other hand, perhaps the lady herself would like to have left one last mystery behind her. . .

...me to a well furnished bedroom, and a large verandah with a good table and other furniture, which... a glance would be most useful to me. Herr Von Essen was charming, and saying that he too was int... in Entomology, stayed on, and helped me very appreciably, with unpacking and arranging th... ...e, I had brought with me from Buea, quite a lot I had too, more than 100, mostly Mylothris, of w... ...ere seemed to be several different species in this country, and in which I was getting grea... ...terested. Then by and by, after Mr. Von Essen had gone, Herr Rein returned, and cam... ...see me; anything but typical German in his appearance—a spare man, rather below... ...height, but with that air of distinction, that is solely and entirely the outcome of good bre... ...was very dark, with a small, black mustache, which did little to hide an exceptionally ug... ...outh, the ugliness of which however was entirely redeemed by an exceptionally charm... ...ile; and in that first moment of meeting with this man, the impression I received was that he... ...greeably surprised, and rather pleased with my appearance, as he looked his approva... ...ke a flash it was gone, forgotten! I only recalled that look, amongst other things, some t... ...ter; merely thinking at the time that perhaps Dr. Graf had been far from flattering i... ...escription he had given of me. Be that as it may, I soon found I had come to a very deli... ...lace; Herr Rein not only would not hear of my paying him any rent for the rooms, b... ...sisted on my staying there entirely as his guest, dining with him every evening, an... ...irst morning, getting up for his very early breakfast, rather before 6 a.m., but aft... ...persuaded him to allow my cook to provide and prepare a breakfast for me in my own... ...the house, and as I took no luncheon, I would have only the pleasure of dining with him e... ...vening; and this arrangement held good all the time I was there. Two days after my arr... ...eing Sunday, Herr Rein invited me to go with him for a drive in his motor-car, up to Buea, to cal... ...Dr. Graf. He drove himself with the native chauffeur sitting next him, while I was in the ton... ...I enjoyed it immensely, but Dr. Graf was not at home; and while the car was cooling off a bu... ...sked if I should have time to go and get some food-plant for my caterpillars, and I re... ...fancy I was away too long on this little expedition, for when I returned to the car, I think he...